T0140599

Lecture Notes in Networks and Systems

Volume 66

Series editor

Janusz Kacprzyk, Polish Academy of Sciences, Systems Research Institute, Warsaw, Poland
e-mail: kacprzyk@ibspan.waw.pl

The series "Lecture Notes in Networks and Systems" publishes the latest developments in Networks and Systems—quickly, informally and with high quality. Original research reported in proceedings and post-proceedings represents the core of LNNS.

Volumes published in LNNS embrace all aspects and subfields of, as well as new challenges in, Networks and Systems.

The series contains proceedings and edited volumes in systems and networks, spanning the areas of Cyber-Physical Systems, Autonomous Systems, Sensor Networks, Control Systems, Energy Systems, Automotive Systems, Biological Systems, Vehicular Networking and Connected Vehicles, Aerospace Systems, Automation, Manufacturing, Smart Grids, Nonlinear Systems, Power Systems, Robotics, Social Systems, Economic Systems and other. Of particular value to both the contributors and the readership are the short publication timeframe and the world-wide distribution and exposure which enable both a wide and rapid dissemination of research output.

The series covers the theory, applications, and perspectives on the state of the art and future developments relevant to systems and networks, decision making, control, complex processes and related areas, as embedded in the fields of interdisciplinary and applied sciences, engineering, computer science, physics, economics, social, and life sciences, as well as the paradigms and methodologies behind them.

** Indexing: The books of this series are submitted to ISI Proceedings, SCOPUS, Google Scholar and Springerlink **

Advisory Board

More information about this series at http://www.springer.com/series/15179

Faddoul Khoukhi · Mohamed Bahaj ·
Mostafa Ezziyyani

Editors

Smart Data and Computational Intelligence

Proceedings of the International Conference
on Advanced Information Technology,
Services and Systems (AIT2S-18) Held
on October 17–18, 2018 in Mohammedia

 Springer

Editors
Faddoul Khoukhi
Faculty of Sciences and Technologies
Mohammedia, Morocco

Mohamed Bahaj
Faculty of Sciences and Technologies
Settat, Morocco

Mostafa Ezziyyani
Faculty of Sciences and Technologies
Boukhalef Tangier, Morocco

ISSN 2367-3370   ISSN 2367-3389   (electronic)
Lecture Notes in Networks and Systems
ISBN 978-3-030-11913-3   ISBN 978-3-030-11914-0   (eBook)
https://doi.org/10.1007/978-3-030-11914-0

Library of Congress Control Number: 2018967945

This Springer imprint is published by the registered company Springer Nature Switzerland AG
The registered company address is: Gewerbestrasse 11, 6330 Cham, Switzerland

Preface

The massive introduction of digital technology in all areas of society (Science, Economy, Army, Health, Finance, Commerce, Education, etc) has radically changed our ways of thinking, behavior, communication, and work. Advanced information technology opens up new opportunities for knowledge creation, education, and information dissemination. It profoundly changes the way countries around the world manage their business and economic affairs, administer public life, and design their political commitment. These innovations have become essential in the transformation of our industry, our services, our health, our education, our tourism, our culture but also the issues of sustainable development, consumption, national security, and our whole democracy; AIT2S-2018 is the International Conference on Advanced Information Technology, Services and Systems. It is a forum for sharing knowledge and results in IT theory, methodology, and applications as well as information technology. The purpose of this conference is to gather researchers and practitioners from academia and industry to discuss and exchange new ideas, results, experiences, and work in progress on all aspects of modern information; engineering concepts, systems, and establishing new collaborations in these areas.

This book collects all papers selected in the AIT2S-2018 conference and which are of scientific interest in the areas related to systems, services, and advanced information technology. These papers represent the work and results of research conducted by researchers and practitioners from academia and industry.

This book includes new ideas, reflections, and experiences of researchers, divided into seven areas:

Advances Networking and Sensor Networks: The sensor networks can be used for various application areas. The objective of this part is to communicate recent and project advances in sensor networks evolution, opportunities, and challenges.

Advances in Software Engineering: Application advances in different domains such as big data, hospital management, aviation, and nuclear power with different requirements have propelled software development from small batch programs to large, real-time programs with multimedia capabilities. To cope, software's enabling technologies have undergone tremendous improvement in hardware, communications, operating systems, compilers, databases, programming languages, and user interfaces, among others. The objective of this part is to know some recent work in this area very rich in scientific work.

Multimedia Systems and Information Processing: Research and development efforts in multimedia computing have opened a wide range of potential applications by combining a variety of information sources such as voice graphics, animation images, audio, and video. This part aims to communicate the latest techniques and systems used in the field of multimedia systems and information processing.

E-learning: This part aims to present the latest trends in the field of e-learning and more particularly the adaptation of a course to the profile of a learner.

Advances Natural Language Processing and Applications: The computing techniques used in the natural language processing have for vocation the learning, the understanding, and the production of the contents in human language. The first researches in linguistics concentrated on the automation of the analysis of the linguistic structure of the language, the machine translation, the speech recognition, and the synthesis of the word. The researchers of today perfect and use these tools in real applications, by creating systems of spoken dialogue and translation engines, by exploiting the social media to obtain information on a large number of subject and domains. We present some papers which handle some aspects of this domain in fast evolution.

IR Big Data, IA, and Knowledge Management: Knowledge is the main wealth of an organization, and it is considered a capital which has an assessable and recoverable economic value. The loss of this knowledge or its bad exploitation leads potentially to a failure of the organization. That's why the knowledge management using techniques of the artificial intelligence and big data has emerged in companies as a major stake.

Intelligent Information Systems and Modeling: The modeling of intelligent information systems has become a very exciting topic in the business world today. Information is no longer only seen as an operational resource, but also as a strategic resource for the company. In this section, we present some papers that raise some aspects of information systems modeling.

Faddoul Khoukhi

Organization

Organizing Committee

Faddoul Khoukhi	FSTM
Mohamed Bahaj	FST-Settat
Mostafa Ezziyyani	FST Tanger

Contents

Advances Networking and Sensor Networks

A New Architecture for Enhancing the QoS of IPTV Video Components by Merging the Advantages of LTE and PHB

Mohamed Matoui[✉], Noureddine Moumkine, and Abdellah Adib

Research Team: Networks, Telecommunications and Multimedia,
Department of Computer Sciences, Faculty of Science
and Technology Mohammedia, Mohammedia, Morocco
matoui.mohamed@gmail.com, nmoumkine@yahoo.fr, adibab@gmail.com

Abstract. IMS (IP Multimedia Subsystem) technologies cannot perform its highest QoS (Quality of Ser-vice) approaches as they aren't able to distinguish between the priorities of IPTV (Internet Protocol Television) video components. In this paper, we propose an improved QoS for IPTV in LTE (Long Term Evolution) systems by providing the capability to prioritize the sub traffic according to the network administrator policy. Our approach is based on a new definition of the IPv6 flow label field who has not yet been standardized. The proposed architecture is implemented using Riverbed Modeler software. The outcomes exhibit that IPTV users receive high-resolution video data with a change in quantity in conformity with data priority.

Keywords: IPv6 · Flow label · LTE · QoS · Diffserv · IPTV · IMS

1 Introduction

In a converged environment, characterized by the emergence of "triple play," the design of new telecommunication architectures, more open, has become a necessity. Such openness will prevent all aspects of fixed/mobile heterogeneity. In the search for an inter-service and inter-network matching solution, IP (Internet Protocol) is presented as a fundamental convergence of NGN (Next Generation Network). It gave a converged IP world in which terminals have become increasingly integrated and ubiquitous. One of the main issues facing telecom operators is to provide multimedia clients with personalized and efficient services depending on the operating environment at the time of service provision. End customers access some multimedia services through completely different devices. These are connected via heterogeneous access networks. The reception quality of the IPTV traffic by the final customer differs according to resources of the acquisition device and network performance. To these quality constraints, add the multimedia traffic sensitivity to QoS parameters namely jitter and packet

© Springer Nature Switzerland AG 2019
F. Khoukhi et al. (Eds.): AIT2S 2018, LNNS 66, pp. 3–16, 2019.
https://doi.org/10.1007/978-3-030-11914-0_1

loss. This is prompting us to design solutions to optimize the QoS by acting on the software components of service delivery i.e., those authenticating and marking users. The negotiation of QoS, as well as the current state of the reception of multimedia product with the entity issuing service, will be an asset to the operator on the horizon to provide traffic with an acceptable QoS. The translation of this philosophy of convergence in an industrial reality materialized in IMS which is a system that serves to cover the convergence of mobile, wireless and fixed networks in a typical network architecture where all data types are housed in an all-IP environment. IMS assumes that the operator must control the media consumption through multimedia sessions using SIP (Session Initiation Protocol) to ensure the required QoS and enable inter convergence services. The latest tests showed that IMS technology still suffers from some confinement factors, which includes the non-differentiation between different IPTV video components. The existing IMS-based IPTV infrastructure doesn't take account that the IPTV traffic consists of three sub-components or the sensitivity of the linear television latency. With the aim of controlling and guaranteeing the QoS in IMS infrastructure, several methods have emerged, namely: 3GPP approach (3rd Generation Partnership Project) [23], IETF (Internet Engineering Task Force) approach [18] and SLM&M (Service Level Monitoring & Management) approach [19]. All these three proposals use the DiffServ (Differentiated Services) model for QoS management. A detailed study shows that traffic classification adopted by these approaches suffers from several problems. The classification of traffic uses three classes: data, voice and video. In connection with IPTV, we state that traffic can be separated into three sub-traffics:

- BC (BroadCast) that licenses the transfer of real-time video.
- VoD (Video on Demand) that includes a library that allows the user to select and view a video.
- PVR (Personal Video Recorder) that allow users to record the received stream.

DiffServ model makes the handling of these three types of flows similar. The difference in sensitivity to QoS parameters requires a reclassification between them. Our contribution aims to remedy this problem. A plethora of types of research have been working on enhancing IPTV services QoS. Li and Chen support IPTV mobility over a wireless cellular network using spectrum allocation technique [3]. This offers better IPTV services by preserving a good quality of voice service. In [4], IPTV data troubles such as dropping, blocking and bandwidth usage have been, to a larger extent, resolved using new queue model that regards adaptive modulation and coding. In terms of using Physical Constraint and Load-Aware in wireless LAN, IPTV seamless handover has been accomplished [11]. This technique allows the user to choose the next wireless LAN to access according to its strength, congestion and bit error rate. In [5], the authors supply a thorough guide to standardized and cutting edge quality assessment models. They also single out and characterize parametric quality of experience (QoE) formulas for the majority of renowned service types (i.e., VoIP, video streaming, online video, etc.), indicating the Key Performance Indicators (KPIs) and major configuration parameters per type. Huang et al propose data-driven QoE prediction for IPTV service. Particularly, they determine QoE to evaluate IPTV user experience in data-driven approach from the scratch and establish a personal QoE model bottomed on an Artificial Neural Network (ANN) [6]. In [2], the development of a testbed for evaluating the QoE of 3D video streaming service over LTE is described. Various network conditions, in the testbed, are configured by setting parameters of network emulator rested on the outcomes reaped by a system-level LTE simulator. In [7], Authors provide a new method to reduce tunneling overhead by allowing multimedia content to be delivered from many different Micro data center as well as Mega data center by utilizing their own special addresses to make tunneling to transmit multimedia content. A new cost-efficient wireless architecture, consisting of a mix of wireless access technologies (satellite, Wi-Fi, and LTE/5G millimeter wave (mmWave) overlay connections), for sending live TV services is suggested in [1]. In our previous work [8], we presented in details the new PHB (Per-Hop Behavior) that reclassifies and differentiates IPTV sub traffics by using the IPv6 Flow Label field. The proposed PHB will make possible prioritization of sub traffics according to the applied QoS network policy. We have already applied our approach on a fixed network [8] and then on a mobile one [9]. In this paper, we introduced a presentation of IMS network and IPTV. We also introduced briefly our new algorithm that helps in increasing the QoS of IPTV sub traffics and implemented it in LTE wireless cellular network to improve the data transfer. Our new QoS optimization mechanism has been explained in Sect. 2. That demonstration is dependent on how to prioritize IPTV sub-traffic utilizing the IPv6 Flow Label field and how to produce new classes of services. In Sect. 3, we discuss our implementation network and the scenario studied of the LTE-IMS-Based IPTV by using Riverbed Modeler. Section 4 reveals the outputs analysis of the suggested scenario of the networks methods. Finally, Sect. 5 discusses the conclusion and our prospects in improving that field.

2   Enhancing QoS Utilizing a New Definition of IPv6 Flow Label

The capacity of the network to supply the user needs upon using IPTV service putting into account the principal parameters such as delay, traffic losses, video jitter and quality is the core of the definition of QoS in our network. Two significant QoS models were offered by IETF (Internet Engineering Task Force): IntServ (Integrated Services) and DiffServ [12,17,22]. The distinction between these models is illustrated further [15]. To upgrade QoS for IPTV services, throughout transmission, IPv6 FL (Flow Label) has been adopted along with IMS system.

2.1   IPv6 Flow Label and Quality of Service

IPv6 FL is a 20-bits field just after the Traffic Class field of the IPv6 header. This field can be applied to label packets of another similar packet flow or an accumulation of flows [10]. A number of approaches have been proposed to the IETF to adopt this field in order to enhance QoS on the internet [20]. Some researchers have proposed its use to deliver the bandwidth, delay, and buffer requirements. Other ones have recommended using this field to send the used port number and the transport protocol [24]. Other approaches have been introduced [26], but none of them have been standardized. However, there is a hybrid approach that takes into account the advanced approaches and applies them to DiffServ model [20]. This model has registered the first 3 bits of the IPv6 FL field to show the methods adopted and kept the remaining 17-bits parameter connecting to each particular approach. Table 1 sums up this hybrid approach.

Table 1. The Bit Pattern for the first 3 Bits of Flow Label

Value	Type of the Used Approach
000	Default
001	A random number is used to define the Flow Label
010	Int-Serv
011	Diff-Serv
100	A format that includes the port number and the protocol in the Flow Label is used
101	A new definition explained in [25]
110	Reserved for future use
111	Reserved for future use

2.2 Optimization of IPTV Broadcasting Traffic

To enhance QoS for IPTV services throughout transmission, IPv6 FL has been utilized with IMS system. The IMS-Based IPTV was not limited to the provision of essential services of IPTV, but it extends to other services like 'quadruple play' services and other more advanced ones as Flow Label to allow the user to ask for a unique process for its real-time traffic flow [22]. Variable data rate has been assigned for video traffic to ensure the best appearance for scenes and modulation process [13,16]. But that assignment causes control and same encapsulation problems for video traffic in a DiffServ network due to the difficulty of designing maximum inter-video traffic limit. Also, when serving considerable traffic with a high priority of EF (Expedited Forwarding) PHB (Per-Hop Behavior), DiffServ core routers faces saturation problem. Because of the growth of real-time data traffic waiting for a delay in the queue due to the use of narrow queues assigned to EF PHB technique. It also causes slow filtering of video packets which leads to dropping it. At the level of dropping process, EF packets at the verge of the DiffServ domain will be processed according to their significance in the GOP (GROUPE of Picture) video [21]. The denial priority for PHBs in AF (Assured Forwarding) is frequently executed based on WRED (Weighted Random Early Detection). Loyalty order user Classification has been integrated into IMS-based

IPTV using the eTOM (enhanced Telecom Operation Map) [19]. Using it, network administrator handles the distinction between recipient based packages. For example, if a user is classified as "GOLD," scoring inter users generate another factor to differentiate between the same user with the same classification to affect the credibility of transmission. But in the congestion case, the DiffServ standard will be used by routers to return to the removal process. As mentioned before, IPTV video data stream can be distinguished into three main traffic: BC, VoD, and PVR. So it will be treated the same in best effort especially in case of traffic congestion [14]. That will lead to traffic latency for sensitive video traffic to latency and loss rate. So the need for reclassification mechanism between IPTV packets became necessary to decrease that delay and packet losses for especially "BC" users. To upgrade QoS for IPTV services throughout transmission, IPv6 FL has been adopted along with IMS system. To achieve that demand, we propose new priority suppression PHBs for IPTV traffic that differentiates between user data depend on their priority [8]. This technique depends on mapping DSCP (Differentiated Services Code Point) values in the IPv6 flow label as the ToS (Type of Service) field of the IPV4 header is limited to a byte. That will give us the capability to differentiate between different IPTV traffic using more bits while remaining compatible with the DiffServ approach. The IPv6 flow label field will thus have the following values (Table 2):

Table 2. New IPv6 flow label values

0	1	2	3	4	5	6	7	8	9	0	1	2	3	4	5	6	7	8	9
0	1	1	DSCP							x	y	Reserved for future use							

Up-to-date IPv6 Flow Label Values and as the value of the DSCP field for the EF class is adjusted to 101110, the IPv6 flow label field can be typed (Table 3):

Table 3. New detailed IPv6 flow label values

0	1	2	3	4	5	6	7	8	9	0	1	2	3	4	5	6	7	8	9
0	1	1	1	0	1	1	1	0	x	y	Reserved for future use								

Where x, y are the bits used to differentiate the video Traffic intra-IPTV. The fact that IPTV packets take the same value of the DSCP field, knowing that it uses merely six bits, then we will use the following 10 and 11-bits in the IPv6 FL to a reclassification intra-IPTV. The rest of 9-bits will be registered for future use. We give the name DSCP-FL to the first 11 bits of the IPv6 FL field. These new Flow Label values are mapped to PHBs that are characterized by a high priority, low loss rate, jitter, and latency analogous to that of the current EF PHB. Indeed, three IPTV packets relating successively under BC traffic, VoD and PVR will be subjected to a treatment explained through the algorithm in Fig. 1: In saturation case, low-level priority data will be removed by the DiffServ; in the explained case, it is to be the one whose DSCP-FL field has a value closer to 01110111001 (Table 4).

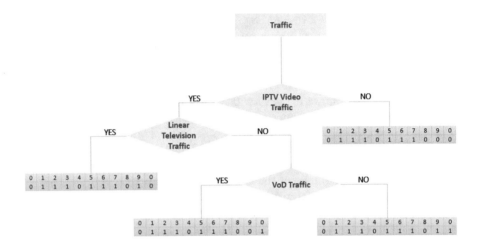

Fig. 1. Our proposed algorithm to differentiate intra-IPTV Traffic

Table 4. Flow label with highest priority level of suppression

0	1	2	3	4	5	6	7	8	9	0	1	2	3	4	5	6	7	8	9	
0	1	1	1	0	1	1	1	0	0	1	Reserved for future use

3   Implementation Scenario

In our regarded network, we intend to raise the data received by BC, VoD and PVR user sequentially and diminish the delay that confronts the BC traffic particularly. Using Riverbed Modeler, we implemented our proposed technique in LTE cellular network. The major idea in the suggested framework architecture is to implement IMS-Based FL IPTV component in a 4G cellular system. A new modulated task application module has been developed as IMS-SIP server does not exist in the Riverbed's modules. User registration in IMS network and session establishment of IPTV services are built in custom application in the proposed framework. In this paper, we contrast between performances of the network in two varied ways. The primary instance without applying the FL QoS based system. The second scenario conforms to applying FL and WFQ (DSCP Based) QoS. Figure 2 introduces three main compositions of the used architecture. The first element comprises the IMS network, the second element

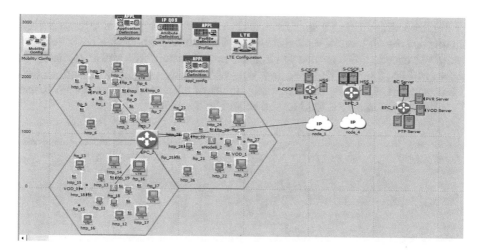

Fig. 2. Moving IPTV user in three different cells

includes three servers that substitute the IPTV data center in charge of delivering different types of multimedia contents (PVR, VoD, and BC). The final one is the personal receiver that acquires data from the sender. As the IPTV users are not the only ones in a 4G network, there are 10 FTP, and 10 HTTP users that transmit data in parallel IPTV users. We should also note that we parallel the conclusions of our suggested scenario upon using and disusing our proposed methods to measure the QoS parameters. The traffic delivered is the same from the three various video servers (PVR, BC, and VoD), high-resolution video, and that after the user perform IMS authentication steps. In that scenario, the users move inside three different cells with the same velocity 100 m/s.

4   Performance Analysis

In this section, we gathered the collected results for our proposed scenario; then we make overall performance analysis. The gathered results are traffic dropped and packet end-to-end delay. We contrast the performance of the three users (BC, VoD, and PVR) in case of using and disusing FL QoS. In this scenario, all the three users BC, VoD and PVR moves inside three different cells with the same speed to make affair comparison between them when applying the suggested technique. Our proposed FL QoS exhibits a high performance for BC user.

4.1   Traffic Dropped

It can be defined as the data missing while sending from the server to the user. This missing data is owing to the congestion of the network and imperfectly data links.

Figure 3 shows that all sources sent the same amount of data, although Fig. 4 shows that the amount of data received by BC user is higher than both VoD user and PVR user as BC user has the highest priority then VoD and PVR users. As shown in Figs. 5 and 6, BC user and VoD user received a higher amount of data when using FL QoS. In contrast, the amount of data received by the PVR user dwindles when using FL QoS as shown in Fig. 7.

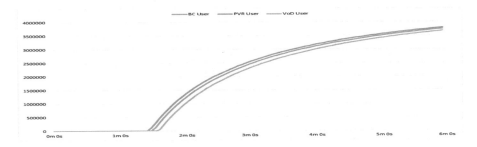

Fig. 3. Traffic sent (bytes/sec)

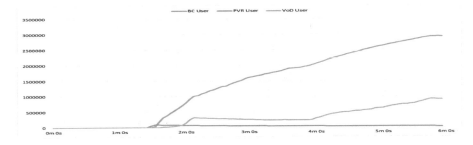

Fig. 4. Traffic received using FL QoS (bytes/sec)

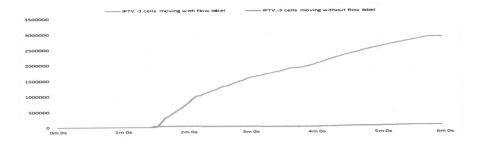

Fig. 5. BC Traffic received (bytes/s)

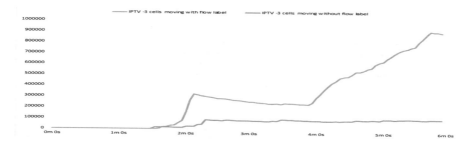

Fig. 6. VoD Traffic received (bytes/s)

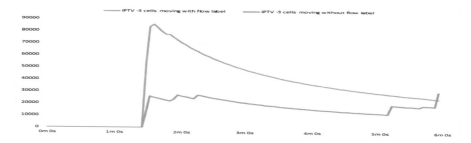

Fig. 7. PVR Traffic received (bytes/s)

4.2 End-to-End Delay

The time taken by the packets to move from the server to the user can be labeled packet end-to-end delay. As shown in Fig. 8, end-to-end delay taken by the BC user is the lowest when using FL QoS. This is because the priority applied in our proposed technique in the proposed scenario. Figure 9 shows that BC packet delay decreases in case of applying our proposed technique. In contrast, Figs. 10 and 11 expresses that the delay of VoD and PVR users raises in case of utilizing the FL QoS technique. The delay of VoD user increases in the small rate while PVR delay rises with high rate.

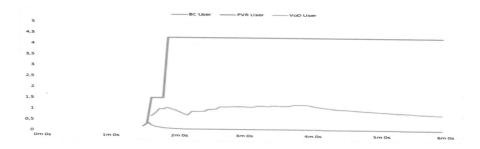

Fig. 8. End-to-End delay (sec)

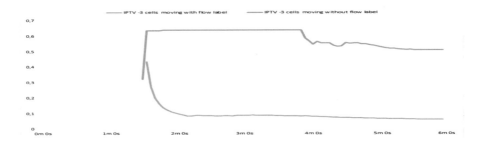

Fig. 9. BC USER End-to-End delay (sec)

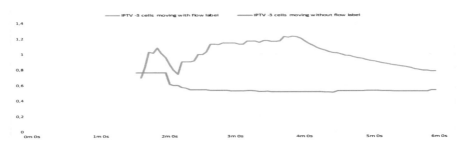

Fig. 10. VoD USER End-to-End delay (sec)

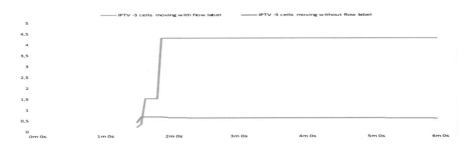

Fig. 11. PVR USER End-to-End delay (sec)

4.2.1   Jitter

It presents the playout buffers size for regular delivery of packets. BC user jitter is the lowest as shown in Fig. 12 when using Flow Label QoS. Figures 13, 14 and 15 represent a comparison of jitter by three users, in the case of using and disusing the Flow Label QoS. BC user Jitter as shown in Fig. 13 is the lower amount of jitter when using our approach. In contrast, the jitter by the VoD and PVR users increases when using Flow Label QoS as in Figs. 14 and 15.

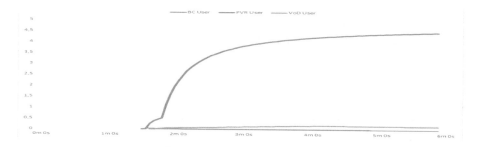

Fig. 12. Jitter (sec)

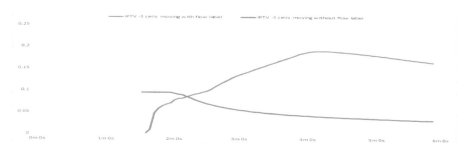

Fig. 13. BC user jitter (sec)

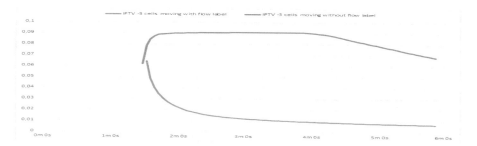

Fig. 14. VoD user jitter (sec)

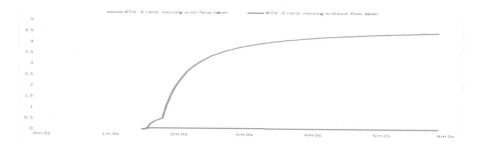

Fig. 15. PVR user jitter (sec)

4.3 Discussion

In this scenario, we notice that in case of network congestion, the router proceeds to the optimization of the delay of the different sub-traffics but with the existence of our mechanism of reclassification and differentiation of the packets, the delay and jitter for PVR user are maximized. As for BC sub-traffic, it has been minimized as a result of our new classification mechanism for BC clients and also because of the sensitivity of this sub-traffics to latency. Our results show that our approach has allowed us to improve the quality of broadcast video to the detriment of VoD and PVR. Such a situation is just a demonstration of the effectiveness of our reclassification approach. In reality, it is up to the administrator to designate the highest priority traffic according to the policy followed by the service provider. The decomposition of IPTV traffic allowed us to set up PHBs specific to IPTV media. The results of the simulations show the added value of our classification algorithm in improving the quality of service of the video.

5 Conclusion and Perspectives

In the recent years, many researchers try to improve the QoS of IPTV services that consider real-time traffic; mainly traffic losses and latency. None of them regards the trouble of classification of IPTV sub traffic and the distinction between the BC, VoD, and PVR packets. To resolve this problem, we suggested a new addressing algorithm that assorts the packets using IPv6 FL field. This algorithm provides a reliable solution to increase QoS of IPTV sub traffic by increasing the priority of BC traffic over VoD and PVR traffics. We also enhance the quality of IPTV services by implementing that method to a 4G cellular system. As LTE system supplies the high bandwidth that supports increasing the quality of conveying data by boosting the amount of transmitted data. We project the performance of this algorithm using an empirical lab as mentioned in Sect. 5. The performance results show that the amount of data received by BC user which had the highest priority is the highest in case of the moving user. Our outcomes demonstrate also that the packet losses, end-to-end delay and jitter went down for BC user, but went up for PVR which explains that our methods work well. We are working on implementing this technique to the next interworking heterogeneous network (LTE-WLAN-WiMAX). In the future, we will work on improving the security issues in IPTV IMS network and solve its related issues.

References

1. Kumar, R., Margolies, R., Jana, R., Liu, Y., Panwar, S.S.: Wilitv, reducing live satellite TV costs using wireless relays. IEEE J. Sel. Areas Commun. 36(2), 233–245 (2018)
2. Solera, M., Toril, M., Palomo, I., Gomez, G., Poncela, J.: A testbed for evaluating video streaming services in LTE. Wirel. Pers. Commun. 98(3), 2753–2773 (2018)

3. Li, M., Chen, L.: Spectrum allocation algorithms for wireless cellular networks supporting mobile IPTV. Comput. Commun. **99**(1), 119–127 (2017)
4. Li, M.: Queueing analysis of unicast IPTV with adaptive modulation and coding in wireless cellular networks. IEEE Trans. Veh. Technol. **66**(10), 9241–9253 (2017)
5. Tsolkas, D., Liotou, E., Passas, N., Merakos, L.: A survey on parametric QoE estimation for popular services. J. Netw. Comput. Appl. **77**, 1–17 (2017)
6. Huang, R., Wei, X., Gao, Y., Lv, C., Mao, J., Bao, Q.: Data-driven QoE prediction for IPTV service. Comput. Commun. **118**, 195–204 (2017)
7. Alsaffar, A., Aazam, M., Hong, C.S., Huh, E.-N.: An architecture of IPTV service based on PVR-Micro data center and PMIPv6 in cloud computing. Multimedia Tools Appl. **76**(20), 21579–21612 (2017)
8. Matoui, M., Moumkine, N., Adib, A.: An IPv6 flow label based approach for IPTV quality of service. In: 2017 International Conference on Wireless Networks and Mobile Communications (WINCOM), pp. 186–192. IEEE (2017)
9. Matoui, M., Moumkine, N., Adib, A.: An IPv6 flow label based approach for mobile IPTV quality of service. In: International Conference on Innovations in Bio-Inspired Computing and Applications, pp. 69–80. Springer (2017)
10. Deering, S., Hinden, R.: Internet Protocol, Version 6 (IPv6) Specification, IETF Internet Draft (2017)
11. Fard, H.S., Rahbar, A.: Physical constraint and load aware seamless handover for IPTV in wireless LANs. Comput. Electr. Eng. **56**, 222–242 (2016)
12. Sambath, K., Abdulrahman, M., Suryani, V.: High quality of service video conferencing over IMS. Int. J. Inf. Educ. Technol. **6**(6), 470 (2016)
13. Farmer, J., Lane, B., Bourg, K., Wang, W.: FTTx Networks, Technology Implementation and Operation. Morgan Kaufmann, Cambridge (2016)
14. Sabry, E.S., Ramadan, R.A., El-Azeem, M.A., ElGouz, H.: Evaluating IPTV network performance using opnet, communication, management and information technology. In: Proceedings of the International Conference on Communication, Management and Information Technology, pp. 377–38. CRC Press (2016)
15. Jiang, S.: Future Wireless and Optical Networks, Networking Modes and Cross-Layer Design. Springer Science & Business Media, Heidelberg (2012)
16. Lloret, J., Canovas, A., Tomas, J., Atenas, M.: A network management algorithm and protocol for improving QoE in mobile IPTV. Comput. Commun. **35**(15), 1855–1870 (2012)
17. 3GPP, 3G TS 23.107 V5.0.0: Quality of Service, Concept, and Architecture (2011). http://www.3gpp.org/ftp/Specs/archive/23_series/23.107/23107-a20.zip
18. Zamora, D.C., Przybysz, H.: Policy and charging control architecture, uS Patent App. 12/934,217 (2011)
19. Raouyane, B., Bellafkih, M., Errais, M., Ranc, D.: WS-composite for management & monitoring IMS network. Int. J. Next-Gener. Comput. **2**(3), 257–270 (2011)
20. Hu, Q., Carpenter, B.: Survey of proposed use cases for the IPv6 flow label, draft-hu-flow-label-cases-03 (2011)
21. Leghroudi, D., Belfkih, M., Moumkine, N., Ramdani, M.: Differentiation intra traffic in the IPTV over IMS context. In: e-Technologies and Networks for Development, pp. 329–336. Springer (2011)
22. Bhattarakosol, P.: Intelligent Quality of Service Technologies and Network Management, Models for Enhancing Communication: Models for Enhancing Communication. IGI Global, Hershey (2010)
23. Siddiqui, M.S., Shaikh, R.A., Hong, C.S.: QoS control in service delivery in IMS. In: 11th International Conference on Advanced Communication Technology, 2009, ICACT 2009, vol. 1, pp. 157–160. IEEE (2009)

24. Prakash, B.: Using the 20 bit flow label field in the IPv6 header to indicate desirable quality of service on the internet, Ph.D. dissertation, University of Colorado (2004)
25. Jee, R., Malhotra, S., Mahaveer, M.: A modified specification for use of the IPv6 flow label for providing an efficient quality of service using a hybrid approach, IPv6 Working Group Internet Draft. Draft-banerjee-flowlabel-ipv6-qos-03. txt, Technical report (2002)
26. Conta, A., Rajahalme, J.: A model for Diffserv use of the ipv6 flow label specification. IETF Internet Draft (2001)

A New IPv6 Security Approach for a Local Network

Ali El Ksimi[(✉)] and Cherkaoui Leghris

L@M, RTM Team, Faculty of Sciences and Technologies,
University Hassan 2 of Casablanca, Mohammedia, Morocco
ali.elksimi@yahoo.fr, cleghris@yahoo.fr

Abstract. Communications using IPv6 over the Internet represent a very minimal rate of all communications. For example, users accessing Google in IPv6 account for only 20.69% of all connections. However, these communications are often present in the majority of local networks, either officially, and after protocol deployment by the administrators, or in the form of "unofficial" traffic using the addresses automatically generated by the nodes of the network. This last case is often neglected by administrators, even if it can be a source of attacks against the network. In this paper, we proposed a new mechanism to secure the messages used in the DAD process, and we compare the results by using two hash functions. Overall, the experimental results show a significant effect in term of execution time.

Keywords: IPv6 · DAD · Security · SHA-1 · SHA-256

1 Introduction

The routing protocol mainly used today for Internet communications is the Internet Protocol (IP). The current IPv4 protocol suffers from many weaknesses and the main problem is the address space. Indeed, the IPv4 addresses are 32 bits long, which represents about 4,3 milliard possible addresses. Following the explosion of network growth Internet and wastage of addresses due to the class structure, the number of IPv4 addresses has become insufficient. Another problem is the saturation of the routing tables in the main routers of the Internet. Even if since 1993, emergency measures have been taken, this only allows delaying the deadline. Also, the Internet Engineering Task Force (IETF) launched work in 1994 to specify the new Internet protocol version IPv6 [17] who will be in replacing IPv4. IPv6 is nowadays ready to support the new Internet trends with more address space and other more important functionalities like neighbor discovery capacity of an IPv6 host.

The Neighbor Discovery (NDP) [18] is the most important part in IPv6; it allows a node to easily integrate into the local network environment, that is, the link on which IPv6 packets are physically transmitted. This protocol makes possible for a host to interact with the equipment connected to the same support (stations and routers). It is important to note that for a given piece of the node, the discovery of the neighbors does not consist in establishing an exhaustive list of the other entire node connected to the link. Indeed, it is only to manage those with whom it dialogues. This protocol performs

F. Khoukhi et al. (Eds.): AIT2S 2018, LNNS 66, pp. 17–26, 2019.
https://doi.org/10.1007/978-3-030-11914-0_2

the following functions: Address Resolution, Neighbor Unreachability Detection, Autoconfiguration, and Redirect Indication using five messages including Router Solicitation, Router Advertisement, Neighbor Solicitation, Neighbor Announcement, and Indication redirection. The IPv6 stateless address Autoconfiguration (SLAAC) [15] is primarily based on the NDP process. This mechanism uses Duplicate address detection (DAD) [1] to verify the uniqueness of the addresses on the same link. However, it is vulnerable to attack, and some solutions have been standardized to minimize this vulnerability, particularly Secure Neighbor Discovery (SEND) [8]. But, they are subject to certain limitations.

In order to secure the DAD process used in IPv6 communications, we develop a new algorithm DAD-Target-Hidden based on SHA-1 [7].

This paper is organized as follows. Section 2 presents a related work to our field. Section 3 describes some IPv6 functionalities, in particular, the DAD process. Section 4 presents the proposed technique. In Sect. 5, we will detail the performance evaluation. Section 6 concludes this paper and addresses some prospects.

2   Related Work

Security attacks in the operation of IPv6, especially DAD process in the NDP, have become one of the interesting research fields. Several proposals have been made by researchers to address security issues in IPv6 DAD.

In [6], the authors have presented a new algorithm for address generation, this mechanism has a minimal computation cost as compared to CGA. Nevertheless, this mechanism uses SHA-1 hash encryption which is vulnerable to collisions attacks.

In [11], the authors have proposed a scheme to secure IPv6 address; this method includes the modifications to the standard RFC 3972 by reducing the granularity factor of the sec from 16 to 8, and replacing RSA with ECC and ECSDSA, using SHA-256 [7] hash function. This method improves the address configuration performance, but it does not eliminate the address conflict.

In [9], the authors have utilized a novel approach for securing IPv6 link-local communication. They have used an alternative approach for the CGA and SEND protocols which still represent a limitation to the security level.

Another approach such as; a secure IPv6 address configuration protocol for vehicular networks [3] was proposed to ensure security in IPv6 without DAD process. However, this method is used only when the distance between a vehicle and its serving AP is one-hop.

In [14], the authors have proposed a new method to secure Neighbor Discovery Protocol in IPv6 which is based on SDN controller to verify the source of NDP packets. However, this method is not efficient because it does not handle the detection of NDP attacks.

Another method was used in [13] to secure the DAD; it is called trust-ND. It is used to detect fake NA messages. However, the experiments show some limits of this method.

In [2], the authors have presented a technique for detecting neighbor solicitation spoofing and advertisement spoofing attacks in IPv6 NDP. However, this method can only detect NS spoofing, NA spoofing, DoS attacks. The disadvantage of this method is that it does not detect other attacks like Duplicate Address Detection attacks.

In [4], the authors have proposed a new method to secure NDP attacks; this method is based on the digital signature. It detects the messages NS and NA spoofing and DOS attacks, router redirection and Duplicate Address Detection, but this mechanism is seen as not complete.

In [10], the authors describe and review some of the fundamental attacks on NDP, prevention mechanisms, and current detection mechanisms for NDP-based attacks.

In this paper, we propose to study and evaluate the security in the NDP within the local network based on IPv6 protocol. Indeed, we propose a new algorithm which could secure the attacks in the DAD process using the hash function SHA-1 and SHA-2 and we compare the execution time of the algorithm with the both of the functions. The experimental results showed that DAD process could be secured by introducing a new field in the NS and NA messages and hide the target address; the hash of the new node's target address. Overall, this method showed a significant effect in term of execution with SHA-1 instead of SHA-256.

3   Duplicate Address Detection (DAD)

3.1   DAD Process

The Neighbor Discovery Protocol mechanism uses ICMPv6 type messages [18]. Under the DAD mechanism, only we are interested in two types of messages, the message Neighbor Solicitation (NS) and the message Neighbor Advertisement (NA). When resolving an address, the message Neighbor Solicitation is used to request the physical address of a node (e.g. MAC address) with it want to communicate by contacting it via his IPv6 address. This message contains a target field that is populated with the node's (IPv6) address that we want to contact. If this target exists, it responds with a message intended for the node that issued the request and contains in one of its fields an option carrying the physical address of this node for the network interface concerned. This association, between the logical and physical addresses, will then be kept in the neighbor cache table.

The DAD mechanism is not infallible, especially if it occurs during the time when several nodes of the same network are temporarily "separated" (loss of connecting or dropping a link between the nodes) and that one or more of the nodes perform a DAD procedure. They can assign the same address without the procedure detects the collision.

For the node, the procedure starts by listening to the multicast group "all-nodes multicast" and the multicast group of the solicited-node ("solicited-node multicast"). The first allows it to receive address resolution requests ("Address Resolution ") for this address and the second will allow it to receive the messages sent by other nodes also making a DAD on this address. In order to listen to these, the node must issue a Multicast Listener Discovery (MLD) [16] request; when a node triggers the DAD procedure, it sends a Neighbor Solicitation message, an ICMPv6 type message.

3.2   The Attack on DAD Process

An attack on the DAD mechanism was identified in [3], the attack is composed as follows: the attacker will deceive the DAD mechanism and make it succeed in one of the two cases where it fails so that the victim cannot claim an address. Since there is a finite number of tries to get an address, the DAD always ends up failing, it's a DoS attack [14]. For the attack to be feasible the attacker must be able to listen on the network any query necessary to perform the DAD procedure (e.g. the NS messages with the unspecified address as the source address are characteristics of the DAD procedure); this implies being able to join the multicast group "Solicited-Node". He then has two choices; he can send an NS message with, as source address, the unspecified address and, as the target address, the address of the victim or an NA message with, as the target address, the "tentative" address of the victim. He can thus prevent the arrival of new nodes having no address yet. The effectiveness of the attack depends strongly on the type of links because it is necessary that the attacker can receive the first NS sent by the victim and that he can answer them. Indeed the attacker must be able to join the multicast group "Solicited-Node", which is not easy in the case of a level 2 point-to-point technology, for example, ADSL.

4   Proposed Technique

In this section, we present the description of our algorithm which makes it possible to secure the target address used in NS message in the DAD process.

4.1   Hash Function

A hash function [12] is a method for characterizing information, a data. By having a sequence of reproducible treatments at an input, it generates a fingerprint to identify the initial data.

A hash function, therefore, takes as input a message of any size, applies a series of transformations and reduces this data. We get at the output a string of hexadecimal characters, the condensed, which summarizes somehow the file.

We define a hash function as an application:

$$\mathbf{h} : \{\mathbf{0}, \mathbf{1}\}^{*} \to \{\mathbf{0}, \mathbf{1}\}^{\mathbf{n}}, n \in \mathbb{N}.$$

The SHA - secure hash algorithm – [5] is a hash algorithm used by certificate authorities to sign certificates and CRL (certificate revocation list). Introduced in 1993 by the NSA with the SHA0, it is used to generate unique condensates (thus for "chopping") of files.

4.2   Hash-TargetAdd-DAD

Hash-TargetAdd-DAD (Hash target address) is a new definition of the ICMPv6 packet (for NS and NA).

Since the standard DAD is not secure, in order to fulfill such security requirement, a "Hash Secure Target" can be applied on NS and NA messages to ensure that only nodes which possess this hash are able to communicate in the IPv6 local network.

In our algorithm, when the node sends an NS message, it assigns the unspecified address to the target address (::), so the target address is hidden. Only nodes that have a specific address can know the target address.

The message format of Hash-TargetAdd-DAD is illustrated in Figs. 1 and 2 Hash-TargetAdd-DAD uses two new message types, namely, $NS_{hash\text{-}targetAdd\text{-}DAD}$ and $NA_{hash\text{-}targetAdd\text{-}DAD}$, and its "Type" fields are 138 and 139, respectively. Compared with the NDP packet, Hash-TargetAdd-DAD adds a new field "Hash_target_64", which stores the last 64 bits of the SHA-1 result.

The hash_target_64 calculation method is illustrated in Figs. 2 and 3 using SHA-256 and SHA-1.

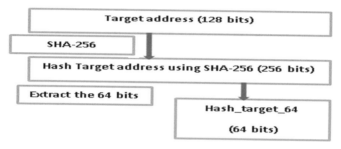

Fig. 1. The message format of Hash-Target-DAD using SHA-256

Fig. 2. The message format of Hash-Target-DAD using SHA-1

Figure 3 shows the $NS_{DAD\text{-}Hide\text{-}Target}$ message.

Target address	:: (hidden address)
Hash_Target_64	H64 (IPv6x)
options	Mac address

Fig. 3. The $NS_{DAD\text{-}Hide\text{-}Target}$

Figure 4 shows the $NA_{DAD\text{-}Hide\text{-}Target}$ message.

Target address	IPv6y
Hash_Target_64	H64 (IPv6y)
options	Mac address

Fig. 4. The $NA_{DAD\text{-}Hide\text{-}Target}$

4.3   Algorithm DAD-Target-Hidden

We present now the two algorithms; the first one represents the sending phase of NS and the receipt of NA, and the second one represents the receipt of NA and its verification.

Algorithm 1. send $NS_{DAD\_Target\_Hidden}$ and receive $NA_{DAD\_Target\_Hidden}$

1: **Input**: address $IPv6_x$
2: **Output**: true: DAD_Target_Hidden success; false: DAD_Target_Hidden fail

3: multicast NS $_{DAD\_Target\_Hidden}$
4: **while** DAD_Target_Hidden timeout ≠ true **do**

5:   receive NA $_{DAD\_Target\_Hidden}$
6:   **if** NA $_{DAD\_Target\_Hidden}$ .Hash_64 == H64 ($IPv6_x$) **then**
7:     **if** NA $_{DAD\_Target\_Hidden}$ .Target address == $IPv6_x$ **then**
8:       **return** false
9:     **else**
10:        **if** H64 (NA$_{DAD\_Target\_Hidden}$ Target address) ≠H64 ($IPv6_x$) **then**
11:          add NA$_{DAD\_Target\_Hidden}$.Src MAC into blacklist
12:        **end if**
13:        drop NA$_{DAD\_Target\_Hidden}$
14:     **end if**
15:   **else**
16:      add NA$_{DAD\_Target\_Hidden}$.Src MAC into blacklist
17:      drop NA$_{DAD\text{-}Target\_Hidden}$
18:   **end if**
19: **end while**
20: **return** true

When another node MN2 receives the $NS_{DAD\_Target\_Hidden}$, it will search in its address pool to find an IP address ($IPv6_Y$) that satisfies the equation:

$$H64\,(IPv6_Y) = NS_{DAD\_Target\_Hidden}.Hash\_64$$

The existence of IPv6$_Y$ indicates a conflicting address and node MN2 needs to send a NA$_{DAD\_Target\_Hidden}$ as a reply to node MN1. The algorithm used in this process is shown in Algorithm 2.

Algorithm 2. receive and verify NS$_{DAD\_Target\_Hidden}$

1: Input: NS$_{DAD\_Target\_Hidden}$
2: Output: true: send NA$_{DAD\_Target\_Hidden}$; false: drop NS$_{DAD\_Target\_Hidden}$
3: receive NS $_{DAD\_Target\_Hidden}$
4: while address pool is not empty do
5:   remove out an IP address as IPv6$_Y$
6:   if H64 (IPv6$_Y$) == NS $_{DAD\_Target\_Hidden}$ Hash_64 then
7:     send out NA $_{DAD\_Target\_Hidden}$
8:     return true
9:   endif
10: end while
11: drop NS $_{DAD\_Target\_Hidden}$
12: return false

4.4   Security Analysis

- Security of "Hash_target_64" field

It supposed that n nodes are present in LAN network, wherein each node has m IPv6 addresses. The length of "Hash_target_64" field is L.
We put l = m x n
The hash collision probability in DAD-Target-Hidden process is:

$$P(Collision) = 1 - \prod_{i=1}^{l} \left(1 - \frac{i}{2^L}\right)$$

Thus, a longer L increases the possibility of node's attack; against a smaller L make the node more secure.

In this work, we set L to 64. In our algorithm DAD-Hide-Target, if 2^8 nodes are present, with each node having 2^{10} IPv6 addresses, then the number of reply messages is:

$$1 - \prod_{i=1}^{2^8 x 2^{10}} \left(1 - \frac{i}{2^{64}}\right) = 1 - e^{-1/29}$$

This value tends to 0, so it is negligible.

5   Performance Evaluation

5.1   Network Topology

The network environment includes a gateway router, an Ethernet switch, a new node (MN1), two existing nodes (MN2 and MN3) and an attacker. Figure 5 shows the network topology. The simulated network is a LAN network using omnet++ simulator.

The network node is less in order to simplify the experimental design and reduce the experimental error. We chose a LAN to simulate our algorithm, since a LAN can have attacks.

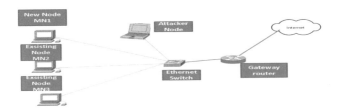

Fig. 5. The network topology

Each node can have several addresses and centralized random address space to increase the probability of address conflict.

5.2   Simulation Results and Evaluation

The simulation results show the following performances:

The number of scenario is 5.

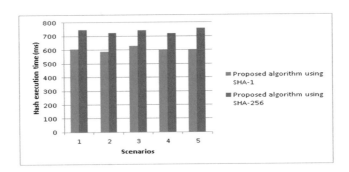

Fig. 6. Hash execution time between the proposed algorithm using SHA-1 and SHA-256

Figure 6 shows the execution time of our algorithm using the functions SHA-1 and SHA-256.

We can see in Fig. 6 that the execution time of our algorithm is faster using SHA-1.

6 Conclusion and Perspectives

In order to ensure that all configured addresses are likely to be unique on a given link, IPv6 nodes execute a duplicate address detection algorithm on the requested IPv6 addresses. Nodes must execute the algorithm before assigning addresses to an interface.

For security reasons, the uniqueness of all addresses must be verified prior to their assignment to an interface. The situation is different for IPv6 addresses created by stateless automatic configuration. The uniqueness of an address is determined primarily by the portion of the address formed from an interface ID. Therefore, if a node has already verified the uniqueness of a link-local address, you do not need to test the additional addresses individually. The addresses must be created from the same interface ID. All manually obtained addresses must be individually tested for their uniqueness. System administrators at some sites believe that the benefits of duplicate address detection are not worth the overhead they use. For these sites, the use of duplicate address detection can be disabled by setting an interface configuration flag.

In this paper, we have developed a new algorithm to secure the DAD process in IPv6 networks. This method is based on the security of NS and NA messages. First, before sending the NS message, the new node uses the hash function SHA-1 or SHA-256 to hash to the target address and extract the last 64 bits and send NS message with the hidden target. When receiving the secure message, the existing nodes compare the hash.

Then, a hash check must be done; so if the hashes are the same, the verification of the IP addresses can be done; otherwise, the message will be deleted.

The results of the simulation show that our algorithm takes less execution time using the SHA-1 hash function instead of SHA-256 which takes longer execution time.

Although IPv6 node communications are limited to NDP and DAD protocols when IPv6 is not officially deployed, there are still attacks that can affect network performance by exploiting only these two protocols as we have been able to see it. Our future work will be focalized on Router discovery messages security.

References

1. Alisherov, F., Kim, T.: Duplicate address detection table in IPv6 mobile networks. In: Chang, C.C., Vasilakos, T., Das, P., Kim, T., Kang, B.H., Khurram Khan, M. (eds.) Advanced Communication and Networking. CCIS, vol 77. Springer, Heidelberg (2010)
2. Barbhuiya, F.A., Bansal, G., Kumar, N., et al.: Detection of neighbor discovery protocol based attacks in IPv6 network. Networking Sci. 2(3–4), 91–113 (2013)
3. Wang, X., Mu, Y., Han, G., Le, D.: A secure IPv6 address configuration protocol for vehicular networks. Wirel. Pers. Commun. 79(1), 721–744 (2014)
4. Hassan, Rosilah, Ahmed, A.S., Othman, N.F.: Enhancing security for IPv6 neighbor discovery protocol using cryptography. Am. J. Appl. Sci. 11(9), 1472–1479 (2014)
5. Aggarwal, S., Aggarwal, S., Aggarwal, K.: A review of comparative study of MD5 and SHA security algorithm. Int. J. Comput. Appl. (0975 – 8887) 104(14), 1–4 (2014)
6. Shah, J.L., Parvez, J.: Optimizing security and address configuration in IPv6 SLAAC. Procedia Comput. Sci. 54, 177–185 (2015)

7. Ibrahim, R.K., et al.: Implementation of secure hash algorithm sha-1 by labview. Int. J. Comput. Sci. Mob. Comput. **4**(3), 61–67 (2015)
8. Moslehpour, M., Khorsandi, S.: A distributed cryptographically generated address computing algorithm for secure neighbor discovery protocol in IPv6. Int. J. Comput. Inf. Eng. **10**(6) (2016)
9. Shah, J.L.: A novel approach for securing IPv6 link local communication. Inf. Secur. J. Global Perspect. (2016). ISSN: 1939-3555
10. Anbar, M., Abdullah, R., Saad, R.M.A., Alomari, E., Alsaleem, S.: Review of security vulnerabilities in the IPv6 neighbor discovery protocol. In: Information Science and Applications (ICISA), pp. 603–612 (2016)
11. Shah, J.L., Parvez, J.: IPv6 cryptographically generated address: analysis and optimization. In: AICTC 2016 Proceedings of the International Conference on Advances in Information Communication Technology & Computing, 12–13 August 2016
12. Abdoun, N., et al.: Secure hash algorithm based on efficient chaotic neural network. IEEE, 04 August 2016
13. Praptodiyono, S., et al.: Improving security of duplicate address detection on IPv6 local network in public area, 31 October 2016, ISSN: 2376-1172
14. Lu, Y., Wang, M., Huang, P.: An SDN-based authentication mechanism for securing neighbor discovery protocol in IPv6. J. Secur. Commun. Networks **2017**, 9 (2017)
15. Gont, F., Cooper, A., Thaler, D., Liu, W.: Recommendation on stable IPv6 interface identifiers. IETF, RFC 8064, February 2017
16. Sridevi: Implementation of multicast routing on IPv4 and IPv6 networks. Int. J. Recent Innov. Trends Comput. Commun., 1455–1467, June 2017. ISSN: 2321-8169
17. Deering, S., Hinden, R.: Internet Protocol, Version 6 (IPv6) specification. IETF, RFC 8200, July 2017
18. Ahmed, A.S., Ahmed, M.S., Hassan, R., Othman, N.E.: IPv6 neighbor discovery protocol specifications, threats and countermeasures: a survey. Electronic, 30 August 2017, ISSN: 2169-3536

Improving IPTV Performance Using IGMP Snooping Protocol

Samiha Moujahid$^{(\boxtimes)}$ and Mohamed Moughit

Mobility and Modeling IR2M Lab., Faculty of Sciences and Techniques,
Hassan 1st University, Settat, Morocco
samiha.moujahid@gmail.com, Mohamed.moughit@gmail.com

Abstract. IPTV is the delivery of television content over IP networks. Who says transmission via IP says the use of several protocols to improve the transmission of the IPTV is at the level of the use of the bandwidth, the loss of the data, the change of the group. In the used protocols one finds IGMP snooping.

IGMP Snooping allows the switch to monitor Internet Group Management Protocol (IGMP) traffic between hosts and multicast routers and uses what it learns to transfer multicast traffic to downstream interfaces that are connected to interested receivers. This helps conserve bandwidth by allowing the switch to send multicast traffic only to interfaces that are connected to devices that want to receive the traffic.

This paper will be dedicated to the study of IGMP Snooping protocol and IPTV performances controlled by IGMP Snooping using OPNET. It shows that when IGMP Snooping is activated in the switch, the channels are transmitted only to subscribers who have requested them, which optimizes the use of the bandwidth more than the other protocols and the quantity of the losses is reduced, also IGMP Snooping is more selective than CGMP and IGMP.

Keywords: IPTV · IGMP snooping · Bandwidth · Group switching · OPNET

1 Introduction

The word IPTV consists of two words IP and TV, IP refers to packets format and addressing schema and TV stands for the communication medium, we are talking here about services offered for television. Therefore, IPTV is a system that provides digital TV services to registered customers for this system, which offers IP over broadband connection and guaranteed quality of service [3, 4].

The formalization of audio and video signals so that they are transmitted on IP packets goes through several steps. The first is the digital analogue conversion because the digital signals are easy to transmit. The second step is the sending of the packets that is made thanks to the intelligent switches that trace the path for the packet, and then go through the gateways that are very intelligent and select the encoders and adjust the protocols and synchronization. The last step is transmission, there are two types of transmission Unicast and Multicast [8]. Unicast causes a bandwidth problem, but multicast does not consume bandwidth [1].

© Springer Nature Switzerland AG 2019
F. Khoukhi et al. (Eds.): AIT2S 2018, LNNS 66, pp. 27–36, 2019.
https://doi.org/10.1007/978-3-030-11914-0_3

The IPTV network contains several television channels. Each channel presents a broadcast group. All TV channels are stored at the head-end and they are sent to the DSLAM. Although there is no demand from the customers, for this, we need a very wide bandwidth [6]. The need for wide bandwidth has led several researchers to focus their studies on the source encoding of multimedia data before transmission [16].

This article shows that if we only stream feeds to the customers who requested them, there will be bandwidth conservation. Thus, the best solution to guarantee a certain selectivity in the case of the change of the group of a user. For these reasons, the article proposes the IGMP Snooping protocol as an effective solution for flow control.

To show these results, OPNET is the most suitable simulator [17, 18].

This paper is divided into four parts, the first part is a presentation of an IPTV network. The second part presents the proposed solution to optimize the use of bandwidth; the third part contains a Discussion of the results obtained. The ultimate is the conclusion from the results as well as the proposals to broaden the research track and, last but not least, the references used to develop the solution.

2   Architecture of IPTV Network

2.1   Hardware of IPTV Network

In an IPTV network, there are five important parts to transmit an end-to-end television signal (Fig. 1):

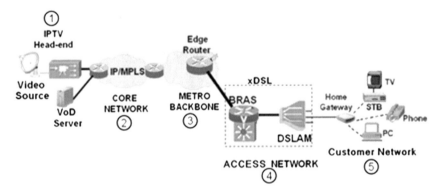

Fig. 1. Architecture of an IPTV network

IPTV Head-end. This is the most important part of the network because it receives the videos from different sources and stores them [1, 12], so it is the entity that aquires, processes, encodes and manages the video content to prepare it for transmission [2, 12].

Core NETWORK. This is the central part of an IPTV network. The central network is the entity that interconnects the metro networks. In general, the core networks use fiber optics as a link to minimize latency between clients and servers as well as improve QoS [2].

METRO BACKBONE. The metro backbone covers an entire region, it is used to insert local TV channels and commercials in, so it provides the service of video on demand to customers in its region. These actions are done through encoders [2].

ACCES NETWORK(xDSL). There are two main elements in this part of the network, ARM and DSLAM.

The BRAS lies between the access and distribution networks and provides access to the Internet and related services to customers. It has several features: Aggregation Point, IGMP Router, Session Termination, Subscriber Management Function, IP Address Assignment, and Policy Management and QoS [1, 2].

In xDSL networks, connections are typically PPP over Ethernet (PPPoE), PPP over ATM (PPPoA), or IP over Ethernet (IPoE) where the connection endpoints are the local equipment of the client (for example, STB) and the arm [2].

This connection is required for user authentication and authorization while connecting to the service. In addition, IGMP messages are sent through this connection [2].

BRAS can monitor individual members by correlating IGMP messages with the PPPoE connection from which they were received.

DSLAM is the point where the subscribers of a region ends, so it can be considered as a point of convergence. DSLAM is connected to the BRAS via an optical fiber, while subscribers are connected to the DSLAM via copper cables. There are three categories of DSLAM: Layer 2 DSLAM, IP DSLAM and IP-aware DSLAM [1].

Customer Network. The customer network provides television, IP phone and Internet services to subscribers. It connects the home's equipment to a broadband service that is provided by network access via a home gateway. This network can also support Voice over IP (VoIP) services [2, 12].

2.2 Protocols of IPTV Network

IGMP. Internet Group Management Protocol (IGMP) is a communication protocol used by hosts and adjacent routers on IPv4 networks to establish multicast group memberships. IGMP is an integral part of IP multicasting. IGMP can be used for one-to-many networking applications, such as video and online games, and enables more efficient use of resources when supporting these types of applications. IGMP is used on IPv4 networks. Multicast management over IPv6 networks is managed by Multicast Listener Discovery (MLD), that uses ICMPv6 messaging as opposed to IGMP blank IP encapsulation [13].

The protocol is relatively simple and is based on 2 messages only:

IGMP membership query: this is the message sent by the router (commonly called DR for designated router) to discover hosts wishing to join a given multicast group. This message is sent at regular intervals to the address 224.0.0.1 with a TTL at 1 (this packet will not leave the local network because of its TTL) [13].

IGMP membership report: packet sent by a given host in order to join a multicast group, this packet has for address of destination the address of the multicast network to which it wants to adhere [13].

This protocol has two major drawbacks:

- First, the host has no way to filter the received packets based on the source address because the source address is not controlled. In fact, this makes it particularly sensitive to DOS attacks.
- The second disadvantage is the latency between when the host no longer wants to be part of the group and when the router actually stops sending data to it. Indeed, no message is expected to request an output, in fact, it will wait until the next issue of IGMP membership query packet by the router. Time to add the time needed for the DR to consider that there is no response and therefore automatically allow an exit.

CGMP. CGMP is a simple protocol, routers are the only devices that produce CGMP messages. The switches only listen to these messages and act accordingly. CGMP uses a well-known destination MAC address (0100.0cdd.dddd) for all its messages. When the switches receive frames with this destination address, they flood them on all their interfaces, so that all switches in the network receive CGMP messages.

In a CGMP message, the two most important elements are:

- Group Destination Address (GDA)
- Unicast Source Address (USA).

Nowadays, the CGMP is rarely used. This is probably because IGMP snooping is supported in the ASIC of any modern switch and unlike CGMP, it is a standard. CGMP owns Cisco and it's hard to find a catalyst IOS switch that supports it.

MBGP. BGP Multiprotocol Extensions (MBGP), sometimes referred to as Multiprotocol BGP or Multicast BGP and defined in IETF RFC 4760, is an extension of Border Gateway Protocol (BGP) that allows to broadcast different types of addresses (called address families) parallel. While standard BGP only supports unicast IPv4 addresses, Multiprotocol BGP supports both IPv4 and IPv6 addresses and supports unicast and multicast variants of each. Multiprotocol BGP allows you to exchange information about the topology of multicast IP routers separately from the topology of normal IPv4 unicast routers. Thus, it allows a multicast routing topology different from the unicast routing topology. Although MBGP allows cross-domain multicast routing information exchange, other protocols such as the Multicast Independent Protocol family are needed to build trees and transmit multicast traffic.

Multiprotocol BGP is also widely deployed in the case of MPLS L3 VPNs, to exchange learned VPN tags for client site routes over the MPLS network, in order to distinguish between different client sites when traffic from other client sites comes to the Peripheral Router Provider (PE router) for routing.

IGMP Snooping. This is a Layer 2 feature that allows you to listen for IGMP traffic by allowing a switch to intercept communications between multicast routers and hosts [14].

As mentioned before, IGMP Snooping allows you to set up multicast groups on networks that do not only work with Cisco hardware. The IGMP snooping principle is to give layer 2 equipment (switch) a layer 3 level [14].

The switch with this feature scans all IGMP traffic between the hosts and the multicast router and performs different operations depending on the type of traffic.

If a host sends a host membership report message to the multicast router, the switch will add the port number of that host to its multicast list. The goal is to send multicast traffic only to this host and not to all hosts as would a switch without IGMP snooping. This process therefore saves bandwidth in networks with multicast groups.

The reverse process occurs when a host sends a leave group to the router, the port number of that host is removed from the switch's multicast table. Although there is a noticeable gain in bandwidth, the switch must scan all IGMP traffic which can cause an overload of the switch processor. This is why it is strongly recommended to implement this feature on routers with a high performance CPU.

This protocol is the subject for this paper.

3 Simulation

OPNET Modeler The Modeler is just one of the many tools from the OPNET Technologies suite. The version used for these experiments was 9.0.A. The engine of the OPNET Modeler is a finite state machine model in combination with an analytical model. Modeler can model protocols, devices and behaviors with about 400 special-purpose modelling functions. The GUI (Graphical User Interface) along with the considerable amount of documentation and study cases that come along with the license are an attractive feature of the Modeler. A number of editors are provided to simplify the different levels of modelling that the network operator requires. Modeler is not open source software though model parameters can be altered, and this can have a significant effect on the simulation accuracy [15, 17].

Figure 2 shows a model of an IPTV network.

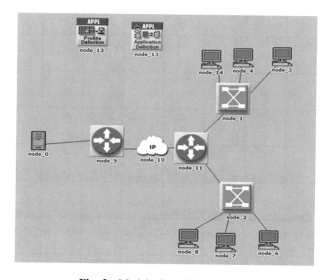

Fig. 2. Model of an IPTV network

The simulation will be divided into two parts, the first part will be dedicated to compare IGMP and CGMP and IGMP Snooping in terms of bandwidth. Whereas,in the second part of the simulation there will be a study on the selectivity of IGMP Snooping in the case of group change by a user.

In the network, there will be two video applications and 2 profiles, a server for storing the televised channels which presents the IPTV head-end part, two routers one which plays the role of DSLAM and the other which plays the role of the meeting. You point, and the switches and televisions that introduce customer network.

There are two multicast groups: one has a multicast address 224.233.24.231 associated with the first video application, and the other has taken 224.233.24.233 as the multicast address associated with the second video application.

4   Result and Discussion

In this part, there is only IGMP that is enabled at the router. So, the switch sends a multicast signal to all members of the multicast group without asking. Figure 3 shows that despite the TV6 is the only one that needs the traffic the switch has sent multicast traffic to all televisions.

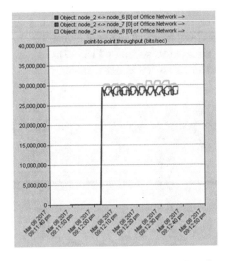

Fig. 3. Received traffic by hosts, only IGMP is active

Now, with the activation of the IGMP only, we test if one of users changes address of the group he can receive the traffic or not. The TV14 will change his group to an other group (Fig. 4).

After this changement, the figure shows that both television TV6 and TV14 receive the same signal (Fig. 5).

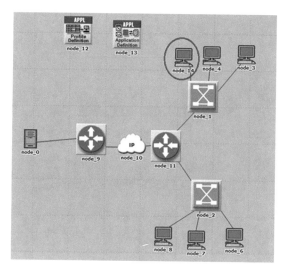

Fig. 4. The host who changed his group

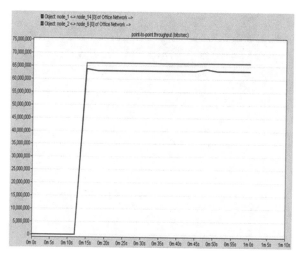

Fig. 5. Traffic received by TV6 and TV14 after changing group user of TV14 when IGMP only activated

In the second scenario, there will be IGMP Snooping activation at switch2, and the TV6 TV user is the only one that requests the traffic. The switch sends traffic only to the user who asked for it (TV6) (Fig. 6).

Also, there will be a comparison between the three IGMP, CGMP and IGMP Snooping protocols. It appears that the IGMP Snooping protocol keeps the bandwidth more than IGMP and CGMP (Fig. 7).

Now, we test the group changing by first activating the CGMP protocol and in the second time the IGMP Snooping protocol. By changing the address of the user group

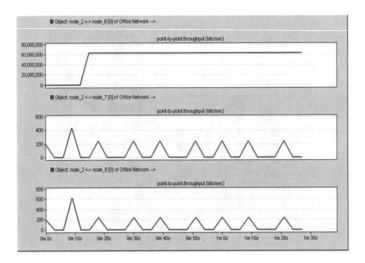

Fig. 6. Traffic received by TV6,TV7, and TV8 when IGMP Snooping is activated

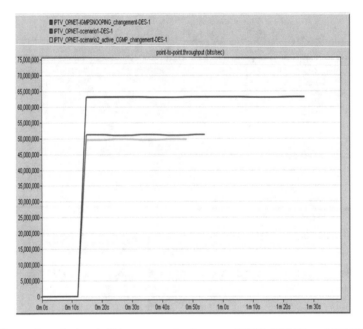

Fig. 7. Comparison the bandwidth conservation between IGMP, CGMP, and IGMP snooping

of TV14 and activating the CGMP protocol, the switches send the signal to the two users that they requested it. On the other hand by activating the IGMP Snooping only the user TV6 receiving the signal (Figs. 8 and 9).

Therefore, the IGMP Snooping protocol is more selective than CGMP.

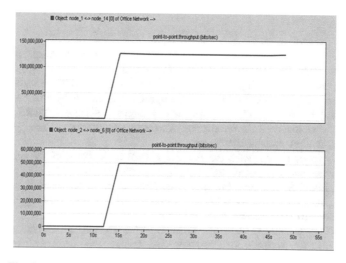

Fig. 8. Traffic received by TV14 and TV6 when CGMP is activated

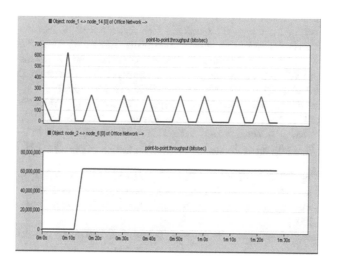

Fig. 9. Traffic received by TV14 and TV6 when IGMP Snooping is activated

5   Conclusion

An IPTV signal is delivered through multicast protocols. In this article, we have focused on two CGMP and IGMP Snooping protocols by comparing the two. We have shown through tests that IGMP snooping brings more benefits for the use and conservation of bandwidth.

References

1. Mohamed, M., Abdelmajid, B., Aicha, S.: A multicast bandwidth saving method. Int. J. Comput. Appl. Technol. (0975–8887) **64**(14) (2013)
2. Abdollahpouri, A., Wolfinger, B.E.: Wired and wireless IPTV access networks: a comparison study. In: IV International Congress on Ultra Modern Telecommunications and Control Systems (2012). Department of Computer Science - TKRN University of Hamburg - Germany 2 University of Kurdistan - Sanandaj - Iran {Abdollahpouri, Wolfinger}@informatik.uni-hamburg.de
3. Malko, S., Uçar, E., Akdeniz, R.: Improving QoE in multicast IPTV systems: channel zapping times. Sci. Res. Essays **7**(35), 3107–3113 (2012)
4. Khider, I., Elfaki, S., Elhassan, M., Siddig, M.: Evaluate the performance of internet protocol television. In: 2011 International Conference on Electrical and Control Engineering (2011). https://doi.org/10.1109/iceceng.2011.6057450
5. Wu, L., Ding, S., Wu, C., Wu, D., Chen, B.: The research and emulation on PIM-SM protocol. In: High Performance Computing and Applications. Lecture Notes in Computer Science, vol. 5938, pp. 465–472 (2010)
6. Huawei Technologies Co., Ltd.: Technical White Paper for HDTV Bearer Networks (2010). Foster. I., Kesselman, C.: The Grid: Blueprint for a New Computing Infrastructure. Morgan Kaufmann, San Francisco (1999)
7. Zhang, J., Wang, Y., Rong, B.: QoS/QoE techniques for IPTV transmissions. In: IEEE International Symposium on Broadband Multimedia System, pp. 1–6 (2009)
8. Telecommunications and Internet converged Services and Protocols for Advanced Networking (TISPAN); NGN integrated IPTV subsystem Architecture Doc. Nb. TS 182 028 Ver. 3.3.1 Ref. RTS/TISPAN-02074- NGNR3, ITU-T (2009)
9. Telecommunications and Internet converged Services and Protocols for Advanced Networking (TISPAN); Service Layer Requirements to integrate NGN Services and IPTV Doc. Nb. TS 181 016 Ver. 3.3.1Ref. RTS/TISPAN-01059-NGN-R3, ITU-T (2009)
10. Minoli, D.: IP Multicast with Applications to IPTV and Mobile DVB-H. Wiley, Hoboken (2008)
11. Hjelm, J.: Why IPTV? Interactivity, Technologies and Services. Wiley, Hoboken (2008)
12. Rahmanian, S.: IPTV Network Infrastructure, Huawei, December 2008
13. Shoaf, S., Bernstein, M.: Introduction to IGMP for IPTV Networks. Juniper Networks Inc, Sunnyvale (2006)
14. Power Connect Application Note #18 February 2004DELL "Understanding IGMP Snooping"
15. Peterson, L.L., Davie, B.S.: Computer Networks a System Approach. Morgan Kaufmann, San Francisco (2003)
16. Majumdar, A., Sachs, D.G.: Multicast and unicast real-time video streaming over wireless LANs. IEEE Trans. Circ. Syst. Video Technol. **12**(6), 524–534 (2002)
17. OPNET Users' Manual, OPNET Architecture, OV.415. http://forums.opnet.com
18. Lucio, G.F., Paredes-Farrera, M., Jammeh, E., Fleury, M., Reed, M.J.: OPNET modeler and Ns-2: comparing the accuracy of network simulators for packet-level analysis using a network testbed. Electronic Systems Engineering Department University of Essex Colchester, Essex C04 3SQ United Kingdom gflore@essex.ac.uk. http://privatewww.essex.ac.uk/~gflore/

Considering the Velocity in the Vertical Handover Network Selection Strategy

Mouad Mansouri$^{(\boxtimes)}$ and Cherkaoui Leghris

L@M Lab, RTM Team, Computer Sciences Department, FST Mohammedia,
Hassan II University, Casablanca, Morocco
mansouri.mouad@yahoo.com

Abstract. The vertical handover is the process of changing the wireless access technology (e.g. from Wi-Fi to 4G/3G). It is an important issue in the next generation of wireless networking. In fact, optimizing this process, especially the discovery and the decision-making steps, allows mobile users to connect to the best available wireless access network anytime, without session ruptures, often caused by wrong vertical handover decisions. This paper addresses the network selection strategy, by using the mobile device's velocity as a constraint during the discovery step and then applying the multi-attribute decision-making methods, to choose in "real-time" the best available wireless network every time. The simulation is inspired by real-life cases, and the results show that our strategy helps to avoid unnecessary handovers.

Keywords: Mobile devices · Network selection · Velocity ·
Vertical handover · Wireless networks

1 Introduction

In a connected world, more and more mobile users (MUs) want to connect wirelessly to the best access technology available anytime, to access the different services they claim. Mobile devices (MDs) are always more sophisticated, they are equipped with multiple wireless access interfaces, allowing them to connect to heterogeneous wireless access networks, either from IEEE or 3GPP groups. Moreover, MDs should connect to the best wireless access network anytime (i.e. always best connected), which leads to vertical handovers (VHs). The VH is the process of changing the point of attachment (PoA) between two heterogeneous wireless access technologies. In opposition, the horizontal handover means the change of the PoA, i.e. base station or access point of homogeneous networks with the same access type [1].

The VH process is divided up to three main stages: The VH initiation, the VH preparation and the VH execution. Figure 1 is a scheme illustrating the different phases and steps of the VH.

IEEE 802.21 or Media Independent Handover (MIH) is the standard that outlines the two first stages [2]. It is implemented between Layer two and Layer

© Springer Nature Switzerland AG 2019
F. Khoukhi et al. (Eds.): AIT2S 2018, LNNS 66, pp. 37–42, 2019.
https://doi.org/10.1007/978-3-030-11914-0_4

Fig. 1. General vertical handover process phases and their steps

three, as a new function, called MIH Function (MIHF). It defines the messages, the commands, and the events that help during the VH. The steps of the VH initiation are:

1. VH discovery: The MD gathers data and link states of the available around wireless networks via MIHF information services (Neighbor map, etc.);
2. VH decision-making (Network selection): In this step, the MD must decide, which available wireless access network is the best, by evaluating different attributes of each network.
3. VH negotiation: After the best available wireless network is selected, the MD must decide whether to continue the VH process to the preparation and execution phases or not.

During the preparation, link layer and IP settings are set up to prepare the communication transfer. Finally comes the handover execution, which is out of the scope of IEEE 802.21, in which handover signaling and context information are broadcasted, and transmission/reception of packets is initiated. In this contribution, we decided to add a velocity condition during the discovery step, that could help in reducing further the number of unnecessary VHs, by pre-eliminating some candidate networks that will only cause session breaks and connectivity loss.

2   Background

The VH decision-making is an important issue in the field of mobile networking. Although MIH standard outlines the steps of the VH process, it does not give well-defined methods, to be used in each step. In [3–6], the authors give an overview of the VH decision process and its steps. They present some selection parameters, used to select the best available wireless network in a heterogeneous environment, and classify the VH strategies based on the techniques used for the VH decision, they list the different criteria used to evaluate a VH decision strategy. In [7] the authors use MADM methods to select the best network, with some modifications in the steps that could improve the decisions quality and offer a better QoS. In [8], the authors present a review and a classification of the most significant MADM algorithms in the context of network selection in heterogeneous wireless networks environments, focusing on the positive and negative points of each decision-making scheme.

Our previous work in [9] is a comparison between different MADM methods, used in VH network selection. We concluded that the Technique for Ordering Preferences by Similarity to the Ideal Solution (TOPSIS), is the best MADM method in this context, using the Fuzzy Analytic Network Process (FANP) to weight the networks' attributes. We made an enhanced decision strategy by considering the battery level every time in [10], by changing the weights vector following the battery level. We tried the fuzzy Gray Relational Analysis (FGRA) in the same context in [11].

The found works considering the MDs' velocity are mainly contributing in VANETs, or other fast-moving conditions, or use location-based information (Using GPS). In [12], the authors propose to optimize the balance between VH attributes and network parameters, based on the speed of the mobile node, and apply GRA to make the VH decisions. In [13], the authors propose a periodic evaluation of available networks, using utility function and cell residence time estimation, based on the mobile terminal's velocity and direction. This strategy helps to avoid unnecessary VHs, but causes some signaling overhead. In [14], the authors introduce a predictive VH decision scheme, considering the direction of the mobile node and the access point load, to cognitively predict the next PoA the MD should connect to. MADM methods and their detailed steps can be found in [9, 15–17].

3   Our Methodology

To evaluate the decisions made in different mobility situations, we designed an application, that will take a dataset containing simulated discovered data as input, and outputs the selected wireless network every time, in different situations from real-life cases, inspired by a movement scenario described in [2]. In this contribution, we consider four heterogeneous decision zones (DZs). A MU is outside (DZ1), goes home and stays for a while (DZ2), before leaving his house to the airport (DZ3). Finally, the MU stays at the airport (DZ4). Moreover, two possibilities are simulated for two different velocities: The first is when the MU is walking with a maximum average velocity of 4.6 Km/h. The second is when the MU moves faster than 4.6 Km/h, case in which it is not very wise to consider the availability of WLANs. The input data is taken from Table 1, every time a wireless technology is available, we variate data from the min to the max value from the table.

We mainly focus in this work, on introducing the velocity constraint in the decision-making step. Our proposal is described in the scheme in Fig. 2. As far as we know, no such decision scheme was proposed in VH decision-making context in the literature. We considered the availability of nine wireless networks, in different decision points: GPRS, EDGE, UMTS, HSxPA, LTE, Wi-Fi a/b/g, Wi-Fi n, Wi-Fi ac and WiMAX. Eight attributes are considered, which are the throughput, the availability, the security, the packet delay, the loss rate, the energy consumption, the cost and the jitter.

Table 1. Possible values of attributes used during the simulation of numeral method

Attributes/ networks	T (Kbps)	Av (%)	S (%)	D (ms)	L ($*10^6$)	EC (1–7)	C (1–7)	J (ms)
GPRS	21.4–171.2	50–100	50	50–70	50–80	2	1	3–20
EDGE	43.2–345.6	40–100	50	20–60	25–70	2	2	3–20
UMTS	144–2000	40–100	60	20–40	15–65	4	4	3–20
HSxPA	14 Mbps	50–100	60	10–50	10–80	4	5	3–20
LTE	10–300 Mbps	40–100	65	10–30	10–40	7	7	3–20
Wi-Fi abg	8–54 Mbps	40–100	60	130–200	30–70	3	1	3–20
Wi-Fi n	72–450 Mbps	30–100	65	100–140	20–60	3	1	3–20
Wi-Fi ac	433–1300 Mbps	50–100	70	90–100	10–40	5	2	3–20
WiMAX	70 Mbps	40–100	60	60–100	10–70	7	5	3–20

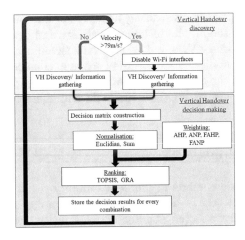

Fig. 2. Flowchart of our network selection algorithm with the velocity constraint

4   Results and Discussion

We run the simulations as described in the previous section, and we recorded the decisions made for two different velocities of the MD. The results are illustrated in Fig. 3, for the best combination of MADM methods: TOPSIS using FANP for weighting. (a) is the record of the decisions made during the first simulation, where the MD's Velocity is low than the threshold, and (b) is the record of the decisions made during the second simulation, where the MD's velocity is higher than the threshold.

The illustrated results in the figures show that our decision strategy allows the MD to choose the best available wireless networks in terms of the delivered QoS. Moreover, the MD does not connect to Wi-Fi networks, when it is moving at an important velocity as we can see in Fig. 3(b). Indeed, our algorithm allows the MD to connect to Wi-Fi when the user is staying home (a), but does not

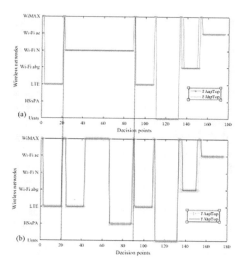

Fig. 3. Records of the decisions made using TOPSIS in the first (a) and the second (b) scenarios

when the user is passing beside his house where the Wi-Fi is available (b). This improvement is very useful because it helps avoiding unnecessary VHs, which will cause session ruptures.

5 Conclusion

In this paper, we propose to consider the velocity of a mobile device, in selecting the best wireless network anytime. The velocity constraint is applied during the network discovery step of the VH process, to pre-eliminate the short coverage wireless networks (WLANs), to avoid handing over them when the MD is moving fast. The results show that our strategy allows this, and makes efficient decisions in terms of the offered QoS. There are still more decision-making methods, that have to be tried in this context to improve the decision quality, and more constraints have to be added in the discovery and negotiation policies, to have more optimal results.

References

1. Mahardhika, G., Ismail, M., Nordin, R.: Multi-criteria vertical handover decision algorithm in heterogeneous wireless network. J. Theor. Appl. Inf. Technol. **54**(2), 339–345 (2013)
2. Gupta, V.: IEEE 802.21 Media Independent Handover. IEEE P802.21 Tutorial (2006)
3. Ferretti, S., Ghini, V., Panzieri, F.: A survey on handover management in mobility architectures. Comput. Netw. **94**, 390–413 (2016)

4. Johal, L.K., Sandhu, A.S.: An overview of vertical handover process and techniques. Indian J. Sci. Technol. **9**(14) (2016)
5. Kassar, M., Kervella, B., Pujolle, G.: An overview of vertical handover decision stratgies in heterogeneous wireless networks. Elsevier Comput. Commun. **31**, 2607–2620 (2008)
6. Bhute, H.A., Karde, P.P., Thakare, V.M.: A vertical handover decision approahes in next generation wireless networks: a survey. Int. J. Mob. Netw. Commun. Telemat. (IJMNCT) **4**(2) (2014)
7. Sadiq, A.S., Fisal, N.B., Ghafoor, K.Z., Lloret, J.: An adaptive handover prediction scheme for seamless mobility based wireless networks. Sci. World J. **14** (2014)
8. Obayiuwana, E., Falowo, O.E.: Network selection in heterogeneous wireless networks using multi-criteria decision-making algorithms: a review. Wirel. Netw., 1–33 (2016)
9. Mansouri, M., Leghris, C.: The use of MADM methods in the vertical handover decision making context. In: International Conference on Wireless Networks and Mobile Communications (WINCOM). IEEE, Rabat (2017)
10. Mansouri, M., Leghris, C.: A battery level aware MADM combination for the vertical handover decision making. In: The 13th International Wireless Communications & Mobile Computing Conference (IWCMC 2017). IEEE, Valencia (2017)
11. Mansouri, M., Leghris, C.: Using fuzzy gray relational analysis in the vertical handover process in wireless networks. In: Networked Systems, NETYS 2017, pp. 396–401. Springer, Marrakech (2017)
12. Khan, M., Han, K.: An optimized network selection and handover triggering scheme for heterogeneous self-organized wireless networks. Math. Probl. Eng. **14** (2014)
13. Yu, H., Yin, J.: Wireless network selection algorithm based on independence of influence factors of QoS. In: 7th International Conference on Computing Communication and Networking Technologies (ICCCNT 2016). IEEE (2016)
14. Shankar, J., Amali, C., Ramachandran, B.: Combined utility and adaptive residence time-based network selection for 4G wireless networks. In: 3rd International Conference on Multimedia Technology (ICMT 2013). IEEE (2013)
15. Saaty, T.L.: Decision making with the analytic hierarchy process. Int. J. Serv. Sci. **1**(1), 83–98 (2008)
16. Saaty, T.L.: Fundamentals of the analytic network process - dependence and feedback in decision-making with a single network. J. Syst. Sci. Syst. Eng. **13**(2), 129–157 (2004)
17. Khademi, N., Mohaymany, A.S., Shahi, J., Zerguini, J.: An algorithm for the analytic network process (ANP) structure design. J. Multi-Criteria Decis. Anal. (2008)

Study of the Impact of Routing on the Performance of IPv4/IPv6 Transition Mechanisms

Khalid El Khadiri$^{(\boxtimes)}$, Ouidad Labouidya, Najib El Kamoun,
and Rachid Hilal

STIC Laboratory, Chouaib Doukkali University, El Jadida, Morocco
{khalid.elkhadiri,labouidya.o,elkamoun.n,
hilal.r}@ucd.ac.ma

Abstract. IPv4/IPv6 transition mechanisms allow sites/hosts to communicate using similar or different formal stacks. Routing protocols constitute an important part of this interconnection. Most of the research work related to the two axes deals with the problems related to the good configurations of the routing protocols and the optimization of their convergence delay. In this paper, we will study the impact of three routing combinations (RIPv2/RIPng, OSPFv2/OSPFv3, and IS-IS) on the performance of three IPv4/IPv6 transition mechanisms (dual stack, manual tunnel, and 6to4 automatic tunnel) by using a real-time application of video conferencing. The study was conducted on the OPNET Modeler network simulator using a number of measurement parameters including delay, delay variation, and packet loss. The results obtained showed that the dual stack mechanism showed the best performance with respect to the tunnelling mechanisms in the case of the three routing combinations envisaged and this for all the measurement parameters.

Keywords: Dual stack · Manual tunnel · 6to4 · RIPng · OSPFv3 · IS-IS · OPNET · Video conferencing

1 Introduction

During the transition phase from IPv4 to IPv6, routing protocols need to adjust to facilitate this transition. In addition, the choice of the best routing protocol among the available is a critical task, which depends, on one hand, on the network requirements and, on the other hand, on the performance parameters of different applications in real time [8]. In principle, routing is a fundamental process for selecting the optimal path from a source node to a destination node in a network.

Most of the routing protocols developed for IPv4 have been updated with minor changes for use with IPv6 [1]. At the configuration level, there are small changes in commands, but the syntax is basically the same. The only difference is the address size, which is of 128 bits in IPv6 instead of 32 bits in IPv4.

In this paper, we will study the impact of three routing combinations (RIPv2/RIPng, OSPFv2/OSPFv3, and IS-IS) on the performance of three IPv4/IPv6 transition mechanisms (dual stack, manual tunnel, and 6to4 automatic tunnel). These mechanisms were

F. Khoukhi et al. (Eds.): AIT2S 2018, LNNS 66, pp. 43–51, 2019.
https://doi.org/10.1007/978-3-030-11914-0_5

evaluated on a simulation network infrastructure under OPNET Modeler using a real-time application of video conferencing. The comparative analysis of the simulation results concerns various parameters such as delay, delay variation, and packet loss. The remaining of the document is organized as follows. Section 2 will present a state of the art of research work performed in this field and our motivation for this research. Simulation scenarios of IPv4/IPv6 transition mechanisms and interior dynamic routing protocols are described in Sect. 3. The results of the simulation and comparative analysis will be discussed in Sect. 4. Conclusions and perspectives will be presented in the final section of this paper.

2 Related Works and Motivation

Many research works have been performed to evaluate the performance of interior dynamic routing protocols and IPv4/IPv6 transition mechanisms for some web applications compared to conventional IP networks.

The authors Sirika and Mahajan [9] performed a comparative study between three dynamic routing protocols that are RIP, EIGRP, and OSPF. This study was based on different criterion such as metrics, routing algorithm, maximum hop number, convergence speed, etc. Consequently, the study mentioned that RIP is suitable for small networks (because of its limited use to 15 hops); the reverse of OSPF, which is intended for large network and EIGRP is a robust protocol but functional only on a Cisco material. In addition, the authors Ashrad and Yousaf [10] studied the performance of the two IPv6 routing protocols, OSPFv3 and EIGRPv6, in terms of convergence time, response time, RTT, overhead tunnel, memory usage, and CPU. As results, they found that OSPFv3 gave better performance than EIGRPv6 regarding most of the simulation parameters.

The authors Kaur and Singh [1], for their part, studied the performance of IS-IS, OSPFv3, and the combination between them (IS-IS_OSPFv3) using the OPNET simulator on two real-time applications (Voice and Video). As results, they found that the IS-IS protocol is better than the others (OSPFv3 and IS-IS_OSPFv3) in terms of video end-to-end delay, OSPFv3 is better in jitter, and IS-IS_OSPFv3 is better in voice end-to-end delay. Another performance evaluation of three IPv6 routing protocols (RIPng, OSPFv3, and EIGRPv6) was performed by the author Al Farizky [6] using streaming video as test traffic on three simulation parameters such as delay, throughput, and packet loss. Consequently, the results showed that EIGRPv6 routing protocol gave better performance than RIPng and OSPFv3.

Other research work studied the performance of the transition mechanisms from IPv4 to IPv6. First of all, the author El Khadiri et al. [4] discussed in a detailed way a comparative study of the mechanisms of transition from IPv4 to IPv6, in which the mechanisms were classified into three families: Dual stack, Tunneling, and Translation, describing for each of them the concerned mechanisms, their working principle, their area of use, their advantages, and their disadvantages. Secondly, they [2, 3] compared the performance of three IPv4/IPv6 transition mechanisms, namely the dual stack, the manual tunnel, and the 6to4 automatic tunnel with two native IPv4 and IPv6 networks. The performance was measured on two real-time applications (VoIP and Video

Conferencing) using two routing protocols, RIPv2 (for the IPv4 infrastructure) and RIPng (for the IPv6 sites), on five simulation parameters such as end-to-end delay, delay variation, jitter, MOS, and packet loss. Consequently, the results showed that the dual stack mechanism gave better performance than the tunneling mechanisms.

Another study was carried out by the author Kalwar et al. [7] whose aim was on the one hand to examine the problems encountered during the process of transition from IPv4 to IPv6 and on the other hand to find the mechanism that can provide transition unnoticeable to final users so that they can use the IPv6 services. The study was performed on GNS3 and Wireshark using ICMP traffic as test traffic. Consequently, the study concluded that the dual stack is the mechanism that provides an unnoticeable transition.

Although researchers have separately studied and evaluated the performance of interior dynamic routing protocols and the IPv4/IPv6 transition mechanisms on many parameters using different traffics and applications, according to our research, no scientific work has studied the impact of routing on the performance of transition mechanisms from IPv4 to IPv6 and the applications transported through these mechanisms. It was a motivation for us to carry out this work under the OPNET Modeler network simulator by studying the impact of three routing combinations (RIPv2/RIPng, OSPFv2/OSPFv3 and IS-IS) on the performance of three IP transition mechanisms (dual stack, manual tunnel and 6to4 automatic tunnel) using a real-time video conferencing application in terms of delay, delay variation and packet loss.

3 Simulation Environment

3.1 Network Topology and Scenarios for Simulation

OPNET Modeler version 14.5 was used to create the simulation scenarios. The network topology represented in Fig. 1 below was used and configured in three different scenarios (dual-stack network, manual tunnel, and 6to4 automatic tunnel) to allow two IPv6 sites (the first is at Larache and the second at Oujda) to communicate with each other through an IPv4 infrastructure composed of 9 routers.

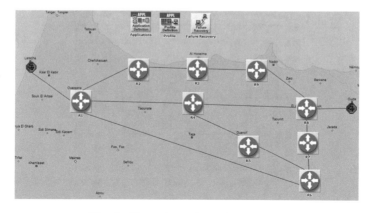

Fig. 1. Network typology of simulation

For routing setup, we have configured the IPv6 versions (RIPng, OSPFv3, and IS-IS) that support the IPv6 on IPv6 sites in parallel with the IPv4 versions (RIPv2, OSPFv2, and IS-IS) that have been configured on the IPv4 infrastructure.

3.2   Elements of Performance Measurement

Three measurement parameters were used during this simulation:

- End-to-end delay: That is the measured time between the moment when a packet is created and sent from a source until it is received at its destination on both IPv6 sites going through the IPv4 infrastructure.
- Delay variation: That is the variation between the unidirectional delay of the selected packets in the same flow of packets [5]. It is based on the difference of end-to-end delay of these selected packets.
- Packet loss: That is the number in percent of lost packets compared to that one of sent packets.

4   Simulation Results and Analysis

4.1   End-to-End Delay

The results obtained in the figures below represent the Video Conferencing Packet End-to-End Delay in milliseconds for the three routing combinations tested.

In terms of the end-to-end delay of video conferencing packets, the results shown in Fig. 2 with RIPv2/RIPng routing combination show that the dual stack gave better performance than the other two IPv4/IPv6 transition mechanisms (manual tunnel and 6to4 automatic tunnel). About an average delay value of 22 ms for the dual stack compared to 26.3 ms and 28.7 ms for the 6to4 automatic tunnel and manual tunnel. This is due to the delay caused by the process of encapsulation and decapsulation in the tunneling mechanisms while in dual stack, the two protocols operate simultaneously without involving either encapsulation or decapsulation.

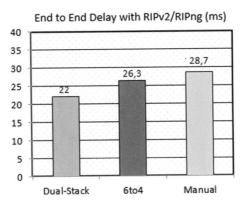

Fig. 2. Video conferencing End to End Delay with RIPv2/RIPng

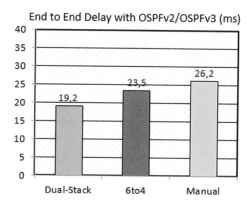

Fig. 3. Video conferencing End to End Delay with OSPFv2/OSPFv3

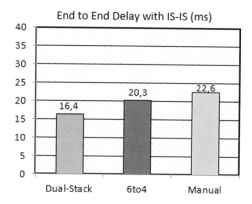

Fig. 4. Video conferencing End to End Delay with IS-IS

In addition, based on the end-to-end delay comparison results of video conferencing packets shown in the figures (Figs. 3 and 4), we notice that the performance of the IPv4/IPv6 transition mechanisms (dual stack, manual tunnel, and 6to4 automatic tunnel) vary in the same way as those of Fig. 2 but with different delay values and depending on the routing combination tested. Indeed, for the three combinations of routing in terms of delay, the dual stack mechanism was better that the mechanisms of tunneling (manual and 6to4). Thus, in the case of the IS-IS protocol, the performance of these mechanisms in terms of delay was better than that with the other combinations RIPv2/RIPng and OSPFv2/OSPFv3. This is justified by the convergence time of the IS-IS protocol which converges rapidly with the topological changes of the network compared to other protocols (RIPv2/RIPng and OSPFv2/OSPFv3).

4.2   Delay Variation

The figures below illustrate the results of Video Conferencing Packet Delay Variation in milliseconds for the three routing combinations tested.

From the results of Fig. 5 in terms of delay variation with the RIPv2/RIPng routing combination, it is clear that the dual stack is more performed than the manual and 6to4 tunneling mechanisms. Indeed, it has a lower value in terms of delay variation of video conferencing packets compared to tunneling mechanisms.

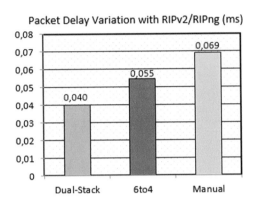

Fig. 5. Video conferencing packet delay variation with RIPv2/RIPng

Also, in the figures (Figs. 6 and 7), in the case of routing combinations (OSPFv2/OSPFv3 and IS-IS), we obtained similar results to Fig. 5 (case of the RIPv2/RIPng routing combination) in terms of the same criteria but with different values. Indeed, in terms of delay variation, the performance of the IPv4/IPv6 transition mechanisms (dual stack, manual tunnel and 6to4 automatic tunnel) is better in the case of the IS-IS protocol than with the other routing combinations tested (RIPv2/RIPng and OSPFv2/OSPFv3).

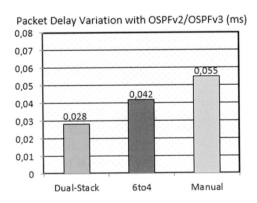

Fig. 6. Video conferencing packet delay variation with OSPFv2/OSPFv3

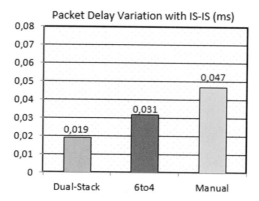

Fig. 7. Video conferencing packet delay variation with IS-IS

4.3   Packet Loss

The results obtained in the figures below represent the Video Conferencing Packet Loss Rate for the three routing combinations tested. According to Fig. 8, in the case of the routing combination (RIPv2/RIPng), the results show that the packet loss rate of the manual and 6to4 tunneling mechanisms is higher than that of the dual stack. About a loss rate of 4.3% and 5% for the 6to4 and manual tunneling mechanisms respectively compared to 3.2% for the dual stack. This is due to the operations of encapsulation/decapsulation of IPv6 packets encapsulated in IPv4 packets by tunneling mechanisms.

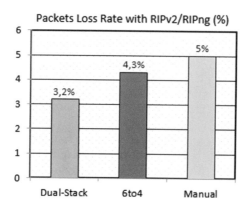

Fig. 8. Video conferencing packets loss rate with RIPv2/RIPng

In addition, results similar to Fig. 8 (case of the RIPv2/RIPng routing combination) in terms of packet loss are found in Figs. 9 and 10 (case of OSPFv2/OSPFv3 and IS- IS routing combinations) but with different percentages. Indeed, in the case of the IS-IS protocol, the packet loss rate of the IPv4/IPv6 transition mechanisms (dual stack, manual tunnel and 6to4 automatic tunnel) is lower than the case of the other routing combinations tested RIPv2/RIPng and OSPFv2/ OSPFv3.

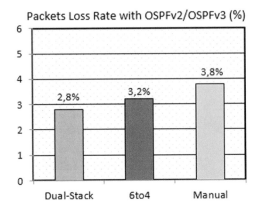

Fig. 9. Video conferencing packets loss rate with OSPFv2/OSPFv3

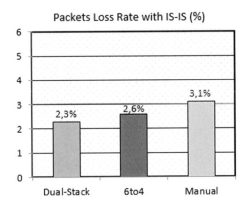

Fig. 10. Video conferencing packets loss rate with IS-IS

5 Conclusion

In this paper, we will study the impact of three routing combinations (RIPv2/RIPng, OSPFv2/OSPFv3, and IS-IS) on the performance of three IPv4/IPv6 transition mechanisms (dual stack, manual tunnel, and 6to4 automatic tunnel). The study was conducted on a real-time application of video conferencing in terms of three measurement parameters, namely delay, delay variation, and packet loss.

The obtained results, for the 3 routing combinations tested for all the measurement parameters envisaged, the performance of the transition mechanisms studied varies in the same way but with different values. Indeed, the dual stack mechanism has shown better performance vis-à-vis the tunneling mechanisms (manual tunnel and 6to4 automatic tunnel). This is explained by the fact that in dual stack, the two protocols operate simultaneously without encapsulation/decapsulation. In addition, this mechanism requires that all network devices support both IPv4 and IPv6 protocols.

Moreover, we also noted by comparing the different routing combinations used between them (the IP transition mechanisms studied), we noticed that the IS-IS routing protocol presents very interesting performances compared to those of the other RIPv2/RIPng and OSPFv2/OSPFv3 protocols for all the measurement criteria envisaged. This is due to the routing process and especially to the convergence time of the protocol which adapts quickly to the topological changes of the network. The goal of this work was to enable customers who want to communicate in IP transition environments to choose the right combination of routing protocols that suits them.

References

1. Kaur, J., Singh, P.: Comparative study of OSPFv 3 IS-IS and OSPFv3_IS-IS protocols using OPNET. IJARCET **3**(8), 2656–2662 (2014)
2. El Khadiri, K., Labouidya, O., Elkamoun, N., Hilal, R.: Performance analysis of video conferencing over various IPv4/IPv6 transition mechanisms. IJCSNS **18**(7), 83–88 (2018)
3. El Khadiri, K., Labouidya, O., Elkamoun,N., Hilal, R.: Performance evaluation of IPv4/IPv6 transition mechanisms for real-time applications using OPNET modeler. Int. J. Adv. Comput. Sci. Appl. IJACSA, Apr. (2018)
4. El Khadiri, K., Labouidya, O.: Etude comparative des mécanismes de transition de l'IPv4 à l'IPv6. Revue Méditerranéenne des Télécommunications, **7**(1) (2017)
5. Neupane, K., Kulgachev, V., Elam, A., Vasireddy, S.H., Jasani, H.: Measuring the performance of VoIP over Wireless LAN. In: Proceedings of the 2011 Conference on Information Technology Education, pp. 269–274 (2011)
6. Al Farizky, R.F.: Routing protocol RIPng, OSPFv3, and EIGRP on IPv6 for video streaming services. In: 2017 5th International Conference on Cyber and IT Service Management (CITSM), pp. 1–6 (2017)
7. Kalwar, S., Bohra, N., Memon, A.A.: A survey of transition mechanisms from IPv4 to IPv6—Simulated test bed and analysis. In: 2015 Third International Conference on Digital Information, Networking, and Wireless Communications (DINWC), pp. 30–34 (2015)
8. Samaan, S.S.: Performance evaluation of RIPng, EIGRPv6 and OSPFv3 for real time applications. J. Eng. **24**(1), 111–122 (2018)
9. Sirika, S., Mahajan, S.: Survey on dynamic routing protocols. Int. J. Eng. Res. Technol. **5**(1), 10–14 (2016)
10. Ashraf, Z., Yousaf, M.: Optimized routing information exchange in hybrid IPv4-IPv6 network using OSPFV3 & EIGRPv6. Int. J. Adv. Comput. Sci. Appl. **8**(4), 220–229 (2017)

Advances in Software Engineering

An Ensemble of Optimal Trees for Software Development Effort Estimation

Zakrani Abdelali[1]([✉]), Moutachaouik Hicham[1], and Namir Abdelwahed[2]

[1] ENSAM, 150 Boulevard Nile, Casablanca, Morocco
{abdelali.zakrani,hicham.moutachaouik}@univh2c.ma
[2] Faculté des Sciences Ben M'sik, Bd Driss El Harti, Casablanca, Morocco
a.namir@yahoo.fr

Abstract. Accurate estimation of software development effort plays a pivotal role in managing and controlling the software development projects more efficiently and effectively. Several software development effort estimation (SDEE) models have been proposed in the literature including machine learning techniques. However, none of these models proved to be powerful in all situation and their performance varies from one dataset to another. To overcome the weaknesses of single estimation techniques, the ensemble methods have been recently employed and evaluated in SDEE. In this paper, we have developed an ensemble of optimal trees for software development effort estimation. We have conducted an empirical study to evaluate and compare the performance of this optimal trees ensemble using five popular datasets and the 30% hold-out validation method. The results show that the proposed ensemble outperforms regression trees and random forest models in terms of MMRE, MdMRE and Pred(0.25) in all datasets used in this paper.

Keywords: Software effort estimation · Optimal trees ensemble ·
Random forest · Regression trees · Accuracy evaluation

1 Introduction

Software development effort estimation (SDEE) is a major area of interest within the field of software project management. An accurate estimate of the software effort plays a central role in delivering projects on time and within budget. Over the last three decades, there has been an important increase in studies dealing with SDEE using a large number of techniques that aimed to improve the accuracy of the estimate and to understand the process used to generate theses estimates. Jorgensen and Shepperd conducted a systematic literature review (SLR) in which they identified up to 11 estimation approaches proposed in 304 selected journal papers [27]. These approaches are based on different techniques varying from expert judgment [25,26] and statistical analysis of historical project data [6,9,28] to artificial intelligence tools [2,14,19,46].

© Springer Nature Switzerland AG 2019
F. Khoukhi et al. (Eds.): AIT2S 2018, LNNS 66, pp. 55–68, 2019.
https://doi.org/10.1007/978-3-030-11914-0_6

Recently, there has been increased emphasis on the use of machine learning (ML) techniques to model the complex relationship between effort and software attributes, especially when this relationship is not linear and doesn't seem to have any predetermined form. Within this context, Wen *et al.* carried out an extensive literature search for relevant studies published in the period 1991–2010 and selected 84 primary empirical studies [44]. They found that eight types of ML techniques have been employed in SDEE models: case-based reasoning (CBR), artificial neural networks (ANN), decision trees (DT), Bayesian networks (BN), support vector regression (SVR), genetic algorithms (GA), genetic programming (GP), and association rules (AR). Among them, CBR, ANN, and DT are most frequently used techniques. Their review also showed that the overall estimation accuracy of most ML models is close to the acceptable level in terms of MMRE and Pred(0.25) and better than that of non-ML models. Nevertheless, each ML technique has its own strength and weakness and the performance of any model depends mainly on the characteristics of the dataset used to construct the model (dataset size, outliers, categorical features and missing values).

In order to overcome the weaknesses of single estimation techniques for prediction tasks, many researchers have investigated the use of ensemble methods in software development effort estimation (SDEE). An ensemble effort estimation (EEE) technique combines several of the single/classical models found in the SDEE literature. MacDonell and Shepperd [35] claimed that combining two or more ML techniques can improve estimation accuracy if no dominant technique can be found. This point of view was approved later by the review done in [27] where it has been revealed that combined model usually generates better estimate than individual model. These findings were also confirmed by Idri *et al.* in their recent review of ensemble effort estimation in which they analyzed 25 studies [23]. However, it has been noted that the number of comparative studies is still insufficient and recommend that researchers should conduct more experiments on ensemble effort estimation techniques and should develop a uniform experimental design [23].

With respect to model combination, regression trees are revealed to be the most used technique to build an ensemble effort estimation model [23]. Most algorithms for obtaining tree ensembles are based on versions of boosting or Random Forest [15,16,37]. Previous work revealed that boosting algorithms exhibit a cyclic behavior of selecting the same tree again and again owing to the way the loss is optimized [41]. At the same time, Random Forest is not based on loss optimization and obtains a more complex and less interpretable model due to the large number of trees included in the forest [12].

In this paper, an ensemble of optimal trees is employed to estimate software development effort. The proposed model aims not only to reduce the number of trees included in the ensemble, but also to improve the estimates accuracy by selecting only the trees having a low error rate. The accuracy estimation of the proposed ensemble is evaluated empirically and compared with regression trees and random forest models using five well-known datasets and the 70–30 hold-out as validation method.

The rest of this paper is structured as follows: Sect. 2 reviews the major previous work related to software development estimation models based on regression tree and ensemble trees. It is followed in Sect. 3 by an overview of the algorithm used to build the ensemble of optimal trees. Section 4 presents a description of historical projects datasets and evaluation criteria employed to evaluate the accuracy of the proposed models. In Sect. 5, we present the experimental design and discuss the results. Finally, the conclusion and future work are given in Sect. 6.

2   Related Work

The following subsections report major works concerning software development effort estimation using regression tree and tree ensemble models.

2.1   Tree-Based SDEE Models

Since its introduction, decision trees have been enjoying increased popularity. The number of applications in the fields of empirical software engineering is growing. In software effort estimation, Selby and Porter generated automatically a large number of decision trees using ID3 algorithm to classify the software modules that had high development effort or faults [40,42]. These decision trees correctly identified 79.3% of the software modules, on the average of across all 9600 trees generated. In [43], the authors compared CARTX, a partial implementation of CART, and backpropagation learning methods to traditional regression approaches. They found that CARTX was competitive with SLIM, COCOMO and Function Points. However, their experiments showed the sensitivity of learning to various aspects of data selection and representation.

In [3], the authors applied fuzzy decision tree on 1000 selected projects data from ISBSG repository R8. Then they extracted a set of association rules to produce mean effort value ranges. The authors claimed the proposed approach provides accurate effort estimation and there is strong evidence that the fuzzy transformation of cost drivers contribute to enhancing the estimation process.

The author in [17] applied a fuzzy version of ID3 on two datasets: COCOMO and Tukutuku. The results obtained indicate the performance of fuzzy ID3 over the crisp version of ID3 in terms of MMRE and Pred(0.25).

In another study [4], Azzeh developed an optimized model tree based on M5P with optimal parameters identified by the Bees algorithm to construct software effort estimation model. The optimized model tree has been validated over eight datasets and evaluated by 3-fold cross-validation method. The results have shown the performance of the optimized model tree over stepwise regression, case-based reasoning and multi-layer perceptron techniques. In [5], the authors employed an evolutionary algorithm to generate a decision tree tailored to a proprietary software effort dataset. The evolutionarily-induced DT statistically outperforms greedily-induced ones (J48, CART and BFTree), as well as traditional logistic regression.

2.2   Tree Ensemble Based SDEE Models

To improve the accuracy of single tree-based model, Elish investigated the use of multiple additive regression trees (MART) in software effort estimation and compared their performance with that of LR, RBF and SVR models on NASA dataset [15]. The MART model outperforms the others in terms of MMRE and Pred(0.25). In [38], the authors designed a Treeboost, also called Stochastic Gradient Boosting, model to predict software effort based on the use case point method. It is to note that the main difference between the Treeboost model and a single decision tree is that the former model consists of a series of trees. Results showed that the Treeboost model outperformed the multiple linear regression model as well as the Use Case Point model in all evaluation criteria adopted in the empirical study.

Despite these promising results, very few studies have assessed the performance of random forest technique in the field of software effort estimation [27] and the only work found in the literature is that of [37]. In this previous work, a comparative study is performed between multiple linear regression (MLR), decision trees (DT) and decision tree forest (DTF). The authors used ISBSG R10 and Desharnais datasets and 10-fold cross-validation method to develop the decision tree forest. The results demonstrate that DTF performs better than MLR and DT in terms of MMRE, MdMRE and Pred(0.25) and the robustness of DTF was confirmed using the non-parametric Mann-Whitney U Test.

3   Optimal Trees Ensemble for SDEE

As the number of trees in random forest is often very large, there has been a significant work done on the problem of minimizing this number not only to reduce computational cost, but also to improve the predictive performance [7,30,32,39]. Since the overall prediction error of a random forest is highly associated with the strength of individual trees and their diversity in the forest. In recent work [30], Khan et al. proposed a further refinement of random forest by proposing a trees selection method on the basis of trees individual accuracy and diversity using unexplained variance.

To this end, we partition the given training data $L = (X, Y)$ randomly into two non-overlapping portions, $L_B = (X_B, Y_B)$ and $L_V = (X_V, Y_V)$. We grow T regression trees on T bootstrap samples from the first portion $L_B = (X_B, Y_B)$. While doing so, select a random sample of $p < d$ features from the entire set of d predictors. This inculcates additional randomness in the trees. Due to bootstrapping, there will be some observations left out of the samples which are called out-of-bag (OOB) observations. These latter take no part in the training of tree. They are used to estimate unexplained variances of each tree grown on a bootstrap sample. Trees are then ranked in ascending order with respect to their unexplained variances and the top ranked M trees are chosen. The selection and combination of trees are carried out as follows:

1. Starting from the two top ranked trees, successive ranked trees are added one by one to see how they perform on the independent validation data, $L_V = (X_V, Y_V)$. This is done until the last M^{th} tree is added.
2. Select tree L_k, $k = \{1, 2, \ldots, M\}$ if its inclusion to the ensemble without the kth tree satisfies the following criteria given by Eq. 1.

$$UnExpVar^{+k} < UnExpVar^{-k} \tag{1}$$

Where $UnExpVar^{-k}$ is the unexplained variance of the ensemble not having the kth tree and $UnExpVar^{+k}$ is the unexplained variance of the ensemble with k^{th} tree included.

The major steps of the coded algorithm is detailed as follows [30]:

Step 1: Take T bootstrap samples from the given portion of the training data $L_B = (X_B, Y_B)$.
Step 2: Grow regression trees on all the bootstrap samples using random forest technique.
Step 3: Choose M trees with the smallest individual prediction error on the training data.
Step 4: Add the M selected trees one by one and select a tree if it improves performance on validation data, $L_V = (X_V, Y_V)$, using unexplained variance performance measures.
Step 5: Combine and allow the trees to average for testing dataset.

4 Data Description and Model Evaluation

This section describes the dataset used to perform the empirical study and presents also the evaluation criteria adopted to compare the estimating capability of the SDEE models.

4.1 Datasets Description

The data used in the present study come from five datasets namely Tukutuku, ISBSG R8, COCOMO, Albrecht and Desharnais. Table 1 displays the summary statistics for these datasets.

– The Tukutuku dataset contains 53 Web projects [36]. Each Web application is described using 9 numerical attributes such as: the number of html or shtml files used, the number of media files and team experience (for more details see Table 1). However, each project volunteered to the Tukutuku database was initially characterized using more than 9 software attributes, but some of them were grouped together. For example, we grouped together the following three attributes: number of new Web pages developed by the team, number of Web pages provided by the customer and the number of Web pages developed by a third party (outsourced) in one attribute reflecting the total number of Web pages in the application (Webpages).

- The ISBSG repository is a multi-organizational dataset containing more than 2,000 projects gathered from different organizations in different countries [24]. Major contributors are in Australia (21%), Japan (20%), and the United States (18%). To decide on the number of software projects, and their descriptions, a data preprocessing study was already conducted by [2], the objective of which was to select data (projects and attributes), in order to retain projects with high quality. The first step of this study was to select only the new development projects with high quality data and using IFPUG counting approach. The second step was concerned by selecting an optimal subset of numerical attributes that are relevant to effort estimation and most appropriate to use as effort drivers in empirical studies.
- The original COCOMO'81 dataset contains 63 software projects [8]. Each project is described by 14 attributes: the software size measured in KDSI (Kilo Delivered Source Instructions) and the remaining 12 attributes are measured on a scale composed of six linguistic values: 'very low', 'low', 'nominal', 'high', 'very high' and 'extra high'. These 13 attributes are related to the software development environment such as the experience of the personnel involved in the software project, the method used in the development and the time and storage constraints imposed on the software. Because the original COCOMO'81 dataset contains only 63 historical software projects and in order to have a robust empirical study, we have artificially generated, from the original COCOMO'81 dataset, three other datasets each one contains 63 software projects (see [22] for more details). The union of the four datasets constitutes the artificial COCOMO'81 dataset that was used in this study.
- The Albrecht dataset [1] is a popular dataset used by many recent studies [20,31,33]. This dataset includes 24 projects developed by third generation languages. Eighteen out of 24 projects were written in COBOL, four were written in PL1, and two were written in DMS languages. There are five independent features: 'Inpcout', 'Outcount', 'Quecount', 'Filcount', and 'SLOC'. The two dependent features are 'Fp' and 'Effort'. The 'Effort' which is recorded in 1,000-person hours is the targeting feature of cost estimation.
- The Desharnais dataset was collected by Desharnais [13]. Despite the fact that this dataset is relatively old, it is one of the large and publicly available datasets. Therefore, it still has been employed by many recent research works, such as [20,31,37]. This dataset includes 81 projects (with nine features) from one Canadian software company. Four of 81 projects contain missing values, so they have been excluded from further investigation. The eight independent features are 'TeamExp', 'ManagerExp', 'Length', 'Language', 'Transactions', 'Entities', 'Envergure', and 'PointsAdjust'. The dependent feature 'Effort' is recorded in 1,000 h.

The descriptive statistics are presented in Table 1. Among these statistics, the 'Skewness' and 'Kurtosis' measures which are used to quantify the degree of non-normality of the features [29]. It can be seen from the data in Table 1 that Albrecht and Desharnais datasets are low order non-normality compared to the rest three datasets.

Table 1. Description statistics of the selected datasets

Dataset	# of software project	# of attributes	Distribution of effort					
			Min	Max	Mean	Median	Skewness	Kurtosis
ISBSG (Release 8)	151	6	24	60 270	5 039	2 449	4.17	21.10
COCOMO	252	13	6	11 400	683.4	98	4.39	20.50
TUKUTUKU	53	9	6	5 000	414.85	105	4.21	20.17
DESHARNAIS	77	8	546	23 940	4 834	3 542	2.04	5.30
ALBRECHT	24	7	0.5	105.20	21.88	11.45	2.30	4.67

4.2 Evaluation Criteria

We employ the following criteria to assess and compare the accuracy of the effort estimation models. A common criterion for the evaluation of effort estimation models is magnitude of relative error (MRE), which is defined as

$$MRE = \frac{|Effort_{actual} - Effort_{estimated}|}{Effort_{actual}} \tag{2}$$

The MRE values are calculated for each project in the dataset, while mean magnitude of relative error (MMRE) computes the average over N projects

$$MMRE = \frac{1}{N}\sum_{i=1}^{N} MRE_i \tag{3}$$

Generally, the acceptable target value for MMRE is 25%. This indicates that on the average, the accuracy of the established estimation models would be less than 25%. Another widely used criterion is the Pred(l) which represents the percentage of MRE that is less than or equal to the value l among all projects. This measure is often used in the literature and is the proportion of the projects for a given level of accuracy. The definition of Pred(l) is given as follows:

$$Pred(l) = \frac{k}{N} \tag{4}$$

Where N is the total number of observations and k is the number of observations whose MRE is less or equal to l. A common value for l is 0.25, which is also used in the present study. The Pred(0.25) represents the percentage of projects whose MRE is less or equal to 0.25. The Pred(0.25) value identifies the effort estimates that are generally accurate whereas the MMRE is fairly conservative with a bias against overestimates [18, 21]. For this reason, MdMRE has been also used as another criterion since it is less sensitive to outliers (Eq. 5).

$$MdMRE = median(MRE_i) \quad i \in \{1, 2, \ldots, N\} \tag{5}$$

4.3   Validation Method

A 30% holdout validation method was used to evaluate the generalization ability of the estimation models. So, the datasets were split randomly into two non-overlapping sets: training set containing 70% of data and testing set composed from 30% of the remaining data. The purpose of holdout evaluation is to test a model on different data to that from which it was learned. This provides less biased estimate of learning performance than all-in evaluation method. Table 2 shows the size of the training and the testing sets.

Table 2. Size of training and testing datasets

Dataset	# of projects in training dataset	# of projects in testing dataset
ISBSG (Release 8)	106	45
COCOMO	176	76
TUKUTUKU	37	16
DESHARNAIS	77	24
ALBRECHT	24	8

5   Experimental Design and Results

This section presents the procedure employed to configure the regression trees, random forest and optimal trees ensemble models and discusses the obtained empirical results.

5.1   Regression Trees Setup

The parameters of the RT model were chosen in such a way that the error is minimal with one exception which is tree pruning. To avoid overfitting, the regression tree was built using 10-fold cross-validation and it was pruned to minimum cross-validation error. The minimum size of the node to split was set at 20 and the maximum depth of tree was set at 30. We chose ANOVA as splitting function since it is a regression tree. The minimum rows allowed in a node was set at 7 as recommended by Breiman [11]. It must be noted that the complexity parameter was set so that any split that does not decrease the overall lack of fit by a factor of 0.01 is not attempted. We used this configuration to generate the 5 regression tree models from the ISBSG, COCOMO, Tukutuku, Albrecht and Desharnais datasets.

5.2   Random Forest Configuration

The random forests method, introduced by Breiman [10], adds an additional layer of randomness to bootstrap aggregating ("bagging") and is found to perform very well compared to many other classifiers. It is robust against overfitting and very user-friendly [34].

The use of random forests in software development effort estimation needs to determine a set of parameters such as: the number of trees constituting the forest (ntree), the number of variables chosen randomly at the level of each node (mtry), the size of the sample 'in bag' (sampsize) and the maximum number of nodes of each tree (maxnodes). In this paper, we use the best configuration as that found in [45] in which a hyperparameter tuning approach was implemented. Table 3 provides an overview of the conducted empirical studies [45].

Table 3. Experimental design of RF models

Dataset	Random Forest parameterization	
	Empirical study 1 varying mtry	Empirical study 2 varying ntree
ISBSG (R8)	from 1 to 5 ntree=300	from 100 to 2000 mtry=5
COCOMO	from 1 to 11 ntree=100	from 100 to 2000 mtry=7
TUKUTUKU	from 1 to 9 ntree=500	from 100 to 1000 mtry=5
ALBRECHT	from 1 to 7 ntree=500	from 100 to 1000 mtry=5
DESHARNAIS	from 1 to 8 ntree=500	from 100 to 1000 mtry=5

5.3   Construction of Optimal Trees Ensemble

As illustrated earlier in Sect. 3, the ensemble of optimal trees is based on random forests. Thus, we used the best setup achieved empirically from Table 3 and we generated 200 random forests from training datasets. Next, our ensemble is built by combining only the top ranked trees from each RF model whose its inclusion decrease the unexplained variance of the ensemble. Table below shows the size of the ensembles obtained in the training phase (Table 4).

5.4   Results and Discussion

Once the three models were trained using training sets, we compared the generalization capability of the proposed ensemble with regression trees and random forest models using the testing sets. The evaluation was based on the MMRE, MdMRE and Pred(0.25) criteria. The complete obtained empirical results are shown in Table 5.

Table 4. Number of trees included in the ensemble of optimal trees

Dataset	Number of trees included
ISBSG (R8)	8
COCOMO	13
TUKUTUKU	1
DESHARNAIS	10
ALBRECHT	1

Table 5. Evaluation of the RT, RF and Ensemble of Optimal Trees models in terms of MMRE, MdMRE and Pred(0.25)(%) over the five datasets using testing sets 30%

Dataset	Regression trees			Random forests			Ensemble of optimal trees		
	MMRE	MdMRE	Pred	MMRE	MdMRE	Pred	MMRE	MdMRE	Pred
ISBSG (R8)	3.71	0,56	26.67	1.17	0,51	33.33	1.02	0.55	40
COCOMO	2.74	0.74	15.79	0.97	0.24	51.31	0.58	0.18	64.47
TUKUTUKU	1.81	0.89	18.75	0.98	0.60	31.25	0.64	0.22	56.25
DESHARNAIS	0.52	0.34	43.48	0.42	0.32	43.48	0.36	0.24	52.17
ALBRECHT	0.97	0.85	28.57	0.73	0.60	42.86	0.3	0.25	57.14

It can be seen from the data in Table 5 that the proposed ensemble of optimal trees performs better than regression trees and random forest models in all datasets. Also, the RF model outperforms RT model in all datasets except Desharnais dataset where their Pred(0.25) are equals.

As shown in Fig. 1, the proposed model made always a smaller MMRE compared to RT and RF models. For the optimal trees ensemble, the lowest MMRE was obtained when using Albrecht whereas the highest MMRE was obtained when using ISBSG R8 dataset. These values of MMREs are not surprising since the Albrecht dataset is exhibiting the lowest non-normality while ISBSG has the highest non-normality according to Kurtosis coefficient (See Table 1). In addition, the large difference between the MMRE values and MdMRE values illustrated in Fig. 2 confirms these findings and demonstrate the presence of significant number of outliers especially in ISBSG R8, COCOMO, and Tukutuku.

Looking at Fig. 3, it is apparent that the ensemble of optimal trees achieved always the highest values of Pred(0.25). i.e. the percentage of the projects whose MRE is less or equal to 0.25. The values of Pred(0.25) achieved a notable increase of 13.5 on average over the five datasets. The best improvement was obtained when using Tukutuku dataset with +25% followed by Albrecht dataset with +14.25.

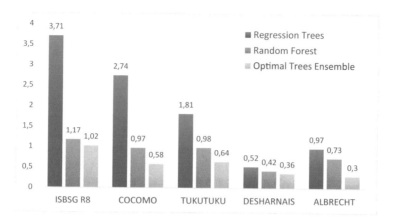

Fig. 1. Comparison of MMRE values for the three SDEE models

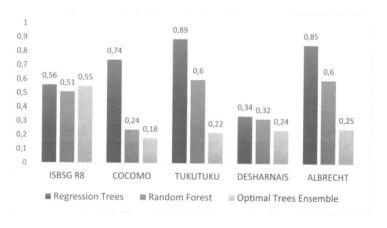

Fig. 2. Comparison of MdMRE values for the three SDEE models

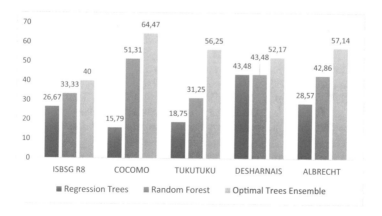

Fig. 3. Comparison of Pred(0.25)(%) values for the three SDEE models

6   Conclusion and Future Work

In this paper, we have empirically investigated the use of a novel method for obtaining an optimal trees ensemble. This latter is built by combining only the top ranked trees from each RF model whose its inclusion decrease the unexplained variance of the ensemble for software development effort estimation. Next, the proposed model was compared to the regression trees and random forest models using 30% hold-out validation method over five datasets namely: COCOMO, ISBSG R8, Tukutuku, Desharnais and Albrecht. The evaluation criteria used were MMRE, MdMRE and Pred(0.25). The results indicated that ensemble of optimal trees surpasses significantly the regression trees and random forest models in terms of all used evaluation criteria. In the light of these empirical results, we can conclude that the ensemble of optimal trees is a promising technique for software development effort estimation. As future work, we are planning to replicate this study using new datasets and employing a leave-one-out validation method.

References

1. Albrecht, A.J., Gaffney, J.E.: Software function, source lines of code, and development effort prediction: a software science validation. IEEE Trans. Softw. Eng. **SE–9**(6), 639–648 (1983)
2. Amazal, F.A., Idri, A., Abran, A.: Software development effort estimation using classical and fuzzy analogy: a cross-validation comparative study. Int. J. Comput. Intell. Appl. **13**(3), 1450013 (2014)
3. Andreou, A.S., Papatheocharous, E.: Software cost estimation using fuzzy decision trees. In: ASE 2008 - 23rd IEEE/ACM International Conference on Automated Software Engineering, pp. 371–374 (2008)
4. Azzeh, M.: Software effort estimation based on optimized model tree. In: 7th International Conference on Predictive Models in Software Engineering, PROMISE 2011, Co-located with ESEM 2011 (2011)
5. Basgalupp, M.P., Barros, R.C., Da Silva, T.S., De Carvalho, A.C.P.L.F.: Software effort prediction: a hyper-heuristic decision-tree based approach. In: 28th Annual ACM Symposium on Applied Computing, SAC 2013, pp. 1109–1116 (2013)
6. Basri, S., Kama, N., Sarkan, H.M., Adli, S., Haneem, F.: An algorithmic-based change effort estimation model for software development. In: Murphy, G.C., Reeves, S., Potanin, A., Dietrich, J. (eds.) 23rd Asia-Pacific Software Engineering Conference, APSEC 2016, pp. 177–184. IEEE Computer Society (2016)
7. Bernard, S., Heutte, L., Adam, S.: On the selection of decision trees in random forests. In: 2009 International Joint Conference on Neural Networks, pp. 302–307. IEEE (2009)
8. Boehm, B.W.: Software Engineering Economics. Prentice Hall PTR, Upper Saddle River (1981)
9. Boehm, B.W., Clark, Horowitz, Brown, Reifer, Chulani, Madachy, R., Steece, B.: Software Cost Estimation with COCOMO II with CDROM. Prentice Hall PTR, Upper Saddle River (2000)
10. Breiman, L.: Random forests. Mach. Learn. **45**(1), 5–32 (2001)

11. Breiman, L., Friedman, J.H., Olshen, R.A., Stone, C.J.: Classification and Regression Trees. Wadsworth, Belmont (1984)
12. Dawer, G., Barbu, A.: Relevant ensemble of trees. CoRR abs/1709.05545 (2017)
13. Desharnais, J.M.: Analyse statistique de la productivitie des projets informatiques a partie de la technique des points de fonction. Master, University of Montreal (1989)
14. Edinson, P., Muthuraj, L.: Performance analysis of FCM based ANFIS and ELMAN neural network in software effort estimation. Int. Arab. J. Inf. Technol. **15**(1), 94–102 (2018)
15. Elish, M.O.: Improved estimation of software project effort using multiple additive regression trees. Expert. Syst. Appl. **36**(7), 10774–10778 (2009)
16. Elish, M.O., Helmy, T., Hussain, M.I.: Empirical study of homogeneous and heterogeneous ensemble models for software development effort estimation. Math. Probl. Eng. **2013** (2013)
17. Elyassami, S., Idri, A.: Applying fuzzy ID3 decision tree for software effort estimation. CoRR abs/1111.0158 (2011)
18. Foss, T., Stensrud, E., Kitchenham, B., Myrtveit, I.: A simulation study of the model evaluation criterion MMRE. IEEE Trans. Softw. Eng. **29**(11), 985–995 (2003)
19. Hosni, M., Idri, A., Nassif, A.B., Abran, A.: Heterogeneous ensembles for software development effort estimation. In: 3rd International Conference on Soft Computing and Machine Intelligence, ISCMI 2016, pp. 174–178. Institute of Electrical and Electronics Engineers Inc. (2016)
20. Idri, A., Abnane, I.: Fuzzy analogy based effort estimation: an empirical comparative study. In: 17th IEEE International Conference on Computer and Information Technology, CIT 2017, pp. 114–121. IEEE Inc. (2017)
21. Idri, A., Abnane, I., Abran, A.: Evaluating Pred(p) and standardized accuracy criteria in software development effort estimation. J. Softw. Evol. Process. **30**(4), e1925 (2018). https://doi.org/10.1002/smr.1925
22. Idri, A., Abran, A., Khoshgoftaar, T.M.: Estimating software project effort by analogy based on linguistic values. In: 8th IEEE Symposium on Software Metrics, METRICS 2002, vol. 2002-January, pp. 21–30. IEEE Computer Society (2002)
23. Idri, A., Hosni, M., Abran, A.: Systematic literature review of ensemble effort estimation. J. Syst. Softw. **118**, 151–175 (2016)
24. ISBSG: International software benchmarking standards group. Data Release 8 Repository, Data Release 8 Repository (2003). http://www.isbsg.org
25. Jorgensen, M.: Practical guidelines for expert-judgment-based software effort estimation. IEEE Softw. **22**(3), 57–63 (2005)
26. Jørgensen, M., Halkjelsvik, T.: The effects of request formats on judgment-based effort estimation. J. Syst. Softw. **83**(1), 29–36 (2010)
27. Jørgensen, M., Shepperd, M.J.: A systematic review of software development cost estimation studies. IEEE Trans. Softw. Eng. **33**(1), 33–53 (2007)
28. Kemerer, C.F.: An empirical validation of software cost estimation models. Commun. ACM **30**(5), 416–429 (1987)
29. Kendall, M., Stuart, A.: The Advanced Theory of Statistics. Vol. 1: Distribution Theory, 4th edn. Griffin, London (1977)
30. Khan, Z., Gul, A., Mahmoud, O., Miftahuddin, M., Perperoglou, A., Adler, W., Lausen, B.: An ensemble of optimal trees for class membership probability estimation. In: Analysis of Large and Complex Data, pp. 395–409. Springer, Cham (2016)

31. Kocaguneli, E., Menzies, T.: Software effort models should be assessed via leave-one-out validation. J. Syst. Softw. **86**(7), 1879–1890 (2013)
32. Latinne, P., Debeir, O., Decaestecker, C.: Limiting the number of trees in random forests. In: Kittler, J., Roli, F. (eds.) Multiple Classifier Systems, pp. 178–187. Springer, Heidelberg (2001)
33. Li, Y.F., Xie, M., Goh, T.N.: A study of the non-linear adjustment for analogy based software cost estimation. Empir. Softw. Eng. **14**(6), 603–643 (2009)
34. Liaw, A., Wiener, M.: Classification and regression by randomforest. R News **2**(3), 18–22 (2002)
35. MacDonell, S.G., Shepperd, M.J.: Combining techniques to optimize effort predictions in software project management. J. Syst. Softw. **66**(2), 91–98 (2003)
36. Mendes, E., Kitchenham, B.: Further comparison of cross-company and within-company effort estimation models for web applications. In: 10th International Symposium on Software Metrics, 2004. Proceedings, pp. 348–357. IEEE (2004)
37. Nassif, A.B., Azzeh, M., Capretz, L.F., Ho, D.: A comparison between decision trees and decision tree forest models for software development effort estimation. In: 2013 3rd International Conference on Communications and Information Technology, ICCIT 2013, pp. 220–224 (2013)
38. Nassif, A.B., Capretz, L.F., Ho, D., Azzeh, M.: A treeboost model for software effort estimation based on use case points. In: 11th IEEE International Conference on Machine Learning and Applications, ICMLA 2012, vol. 2, pp. 314–319 (2012)
39. Oshiro, T.M., Perez, P.S., Baranauskas, J.A.: How many trees in a random forest? In: International Workshop on Machine Learning and Data Mining in Pattern Recognition, pp. 154–168. Springer (2012)
40. Porter, A.A., Selby, R.W.: Evaluating techniques for generating metric-based classification trees. J. Syst. Softw. **12**(3), 209–218 (1990)
41. Rudin, C., Daubechies, I., Schapire, R.E.: The dynamics of AdaBoost: cyclic behavior and convergence of margins. J. Mach. Learn. Res. **5**, 1557–1595 (2004)
42. Selby, R.W., Porter, A.A.: Learning from examples: generation and evaluation of decision trees for software resource analysis. IEEE Trans. Softw. Eng. **14**(12), 1743–1757 (1988)
43. Srinivasan, K., Fisher, D.: Machine learning approaches to estimating software development effort. IEEE Trans. Softw. Eng. **21**(2), 126–137 (1995)
44. Wen, J., Li, S., Lin, Z., Hu, Y., Huang, C.: Systematic literature review of machine learning based software development effort estimation models. Inf. Softw. Technol. **54**(1), 41–59 (2012)
45. Zakrani, A., Hain, M., Namir, A.: Investigating the use of random forests in software effort estimation. In: Second International Conference on Intelligent Computing in Data Sciences (ICDS 2018) (2018)
46. Zakrani, A., Idri, A.: Applying radial basis function neural networks based on fuzzy clustering to estimate web applications effort. Int. Rev. Comput. Softw. **5**(5), 516–524 (2010)

A Multi-objective Model for the Industrial Hazardous Waste Location-Routing Problem

Regragui Hidaya[(⊠)] and Mohammed Karim Benhachmi

Processes and Environmental Engineering Laboratory,
Faculty of Sciences and Techniques of Mohammedia, Mohammedia, Morocco
hidaya.regragui@gmail.com, benhachmikarim@gmail.com

Abstract. The objective of Hazardous Waste Management (HWM) is to ensure safe, efficient and cost-effective collection, transportation, treatment and disposal of Hazardous Waste. The hazardous waste location routing problem aims to answer the following questions: where to open treatment centers and with which technologies, where to open disposal centers, how to route different types of hazardous waste to which of the compatible treatment technologies, and how to route waste residues to disposal centers. The goal of this paper is to adapt an existing model to Moroccan regulation by adding a new risk equity objective for HWM.

Keywords: Location routing problem ·
Industrial hazardous waste management · Multi-objective optimization

1 Introduction

Hazardous waste is a waste that is dangerous or potentially harmful to our health or the environment, it can be ignitable, reactive, corrosive or toxic.

The increasing developments in technology and industry have led to a significant hazardous waste management problem, demanding a more structured and scientific manner of managing hazardous material. For instance, Morocco is currently generating 242 919 tons/year of industrial hazardous waste, this amount is expected to grow to 329 419 tons/year in 2028 [2].

According to Nema and Gupta [7] "Hazardous waste management is collecting, transporting, treating, recycling, and disposing of residue under safe, efficient, and cost-effective manner". The complexity of HWM comes from different aspects: multiple kinds of hazardous waste, multiple kinds of treatment technologies, compatibility issues (i.e. which treatment facility is compatible with a given waste kind) and the uncertainty of hazardous waste generation.

In this paper, we present a hazardous waste location-routing model help decision maker decide on locations of treatment centers utilizing different technologies, routing different types of industrial hazardous wastes to compatible treatment centers, and routing hazardous waste and waste residues to those centers, and locations of disposal centers and routing waste residues there; it could be also a tool for policy evaluation.

© Springer Nature Switzerland AG 2019
F. Khoukhi et al. (Eds.): AIT2S 2018, LNNS 66, pp. 69–77, 2019.
https://doi.org/10.1007/978-3-030-11914-0_7

2 Literature Review

The first effort to combine location and routing for hazardous waste was by Zografos and Samara [11]. They propose a goal-programming model for one type of hazardous waste, optimizes the travel time, transportation risk, and site risk.

Later Jacobs and Warmerdam [4] considers a continuous network flow problem to model the hazardous waste location-routing problem. Their model locates the storage and disposal facilities and determines the routing strategies while minimizing the linear combination of cost and risk in time. An additional consideration in Wyman and Kuby [9] presented a model with three objectives minimizing cost and risk, maximize equity by analyzing the transportation and facility location components of risk and equity separately. Giannikos [3] uses the weighted goal-programming technique to attempt four targets (objectives) minimization of cost, minimization of total perceived risk, equitable distribution of risk among population centers, and equitable distribution of disutility caused by the operation of the treatment facilities.

List and Mirchandani [5] developed a model that consider waste types and has three objectives: minimization of risk, minimization of cost, and maximization of equity. Later, Nema and Gupta [7] introduced two new constraints: waste–waste and waste–technology compatibility constraints. The waste–waste compatibility constraint ensures that a waste is transported or treated only with a compatible waste, and the waste–technology compatibility constraint ensures that a waste is treated only with a compatible technology.

Alumur and Kara [1] developed linear constraints to formulate the waste-technology compatibility constraint (i.e. a waste can be treated using only its compatible technology). Later Samanlioglu [8] added recycling location and routing into their waste management system.

Most of the hazardous waste location-routing literature does not reflect on real-life aspects, such as considering multiple waste types, waste-waste and waste-technology compatibility, and uncertainty of variable cost, risk and generated amount of hazardous waste. The first to incorporate uncertainty into hazardous waste location routing problem was Zhang and Zhao [10] they focused on the fuzzy demand of waste.

3 Problem Description

The objective of the mathematical model is to response questions related to: the locations of treatment centers with different technologies; routing different types of hazardous wastes to compatible treatment centers; the routing hazardous wastes and waste residues to these centers; and the locations of disposal centers and routing waste residues to these centers.

Alumur and Kara [1] described the hazardous waste LRP as follows "Given a transportation network and the set of potential nodes for treatment and disposal facilities, find the location of treatment and disposal centers and the amount of shipped hazardous waste and waste residue, to minimize the total cost and the transportation risk". They proposed a model to treat all the generated hazardous waste and to dispose all the generated waste residues in a safe and cost-effective manner respecting waste-technology constraint.

The model didn't incorporate existing treatment and disposal facilities. Recycling is not either adopted, because they assumed that hazardous waste may not always be suitable for recycling.

Our proposal solves the hazardous waste location routing problem where existing treatment and disposal facilities are taking into account and considering perceived risk to the risk objective.

3.1 Assumptions

- Given a transportation network, the nodes may be a generation node, a transshipment node (a node junction), a potential treatment facility, a potential disposal facility, or a combination of any of the above.
- The potential sites for treatment and disposal facilities have already been identified.
- The costs of transporting one unit of hazardous waste and one unit of waste residue are directly proportional to the network distance used.
- All hazardous wastes are transported with the same type of trucks at the same unit costs (i.e. transportation cost only depends on the distance of transport and the amount of hazardous waste transported)
- The transportation costs of hazardous waste and waste residue may differ, as special trucks or containers may be needed to transport hazardous waste, while waste residue can be transported casually as domestic waste.
- Fixed cost in locating treatment and disposal facilities depend on the treatment technology employed the size of the facility to be located, or other factors.

3.2 Mathematical Model

W The following model is a multi-objective Mixed Integer Programming model for the industrial HWLRP that extends the formulation proposed by Alumur and Kara [1], by adding

- The individual perceived risk.
- Existing treatment and disposal facilities.

$N = (V, A)$ transportation network
$G = \{1, \ldots, g\}$ generation nodes
$T = \{1, \ldots, t\}$ potential treatment nodes
  T' existing treatment nodes, $T' \subset T$
$D = \{1, \ldots, d\}$ potential disposal nodes
  D' existing disposal nodes, $D' \subset D$
$Tr = \{1, \ldots, tr\}$ transshipment nodes
$W = \{1, \ldots, w\}$ hazardous waste types
$Q = \{1, \ldots, q\}$ treatment technologies

Parameters:

$c_{i,j}$ cost of transporting one unit of hazardous waste on link $(i,j) \in A$

$cz_{i,j}$ cost of transporting one unit of waste residue on link $(i,j) \in A$

$fc_{q,i}$ fixed annual cost of opening a treatment technology $q \in Q$ at treatment node $i \in T$

fd_i fixed annual cost of opening a disposal facility at disposal node $i \in D$

$POP_{w,i,j}$ number of people in the bandwidth for HW type $w \in W$ along link $(i,j) \in A$

POP_i number of people in the bandwidth of treatment or disposal facility $i \in (T+D)$ $g_{w,i}$ amount of hazardous waste type $w \in W$ generated at generation node $i \in G$

$\alpha_{w,i}$ recycle percent of hazardous waste type $w \in W$ generated at generation node $i \in G$

$\beta_{w,q}$ recycle percent of hazardous waste type $w \in W$ treated with technology $q \in Q$

$r_{w,q}$ percent mass reduction of hazardous waste type $w \in W$ treated with technology $q \in Q$

$t_{q,i}$ capacity of treatment technology $q \in Q$ at treatment node $i \in T$

$t^m_{q,i}$ minimum amount of hazardous waste required for treatment technology $q \in Q$ at treatment center $i \in T$

dc_i disposal capacity of disposal site $i \in D$

$com_{w,q} = 1$ if waste type $w \in W$ is compatible with technology $q \in Q$; 0 otherwise

Decision variables:

$x_{w,i,j}$ amount of hazardous waste type w transported through link (i, j)

$z_{i,j}$ amount of waste residue transported through link (i, j)

$y_{w,q,i}$ amount of hazardous waste type w to be treated at treatment node i with technology q d_i amount of waste residue to be disposed of at disposal node i

$f_{q,i} = 1$ if treatment technology q is established at treatment node i; 0 otherwise

$dz_i = 1$ if disposal site is established at disposal node i; 0 otherwise

In the model, the non-recycled amount of generated hazardous wastes $((1-\alpha_{w,i})g_{w,i})$ are to be routed $(x_{w,i,j})$ to the compatible treatment technology in the treatment facility $(y_{w,q,i})$ to be located $(f_{q,i})$. After the treatment process, the non-recycled amount of waste residues are to be routed $(z_{i,j})$ to the ultimate disposal facility, which is also to be located (d_i)

$$\text{Minimize} \sum_{(i,j)\in A} \sum_w c_{i,j} x_{w,i,j} + \sum_{(i,j)\in A} cz_{i,j} z_{i,j} + \sum_i \sum_q fc_{q,i} f_{q,i} + \sum_i fd_i dz_i$$

$$\text{Minimize} \sum_{(i,j)\in A} \sum_w POP_{w,i,j} x_{w,i,j} + \sum_{i\in(T+D)} \sum_q \sum_w POP_i y_{w,q,i}$$

s.t.

$$((1 - \alpha_{w,i})g_{w,i}) = \sum_{j:(i,j)\in A} x_{w,i,j} - \sum_{i:(i,j)\in A} x_{w,j,i} + \sum_q y_{w,q,i}, w \in W, i \in V \quad (1)$$

$$\sum_q \sum_w y_{w,q,i}(1 - r_{w,q})(1 - \beta_{w,q}) - d_i = \sum_{j:(i,j)\in A} z_{i,j} - \sum_{i:(i,j)\in A} z_{j,i}, i \in V \quad (2)$$

$$\sum_w y_{w,q,i} \le t_{q,i} f_{q,i}, q \in Q, i \in T \tag{3}$$

$$d_i \le dc_i\, dz_i\,, i \in D \tag{4}$$

$$\sum_w y_{w,q,i} \ge t^m_{q,i} f_{q,i}\,, q \in Q, i \in T \tag{5}$$

$$y_{w,q,i} \le t_{q,i}\, com_{w,q}\,\,, w \in W, q \in Q, i \in T \tag{6}$$

$$\sum_q \sum_w y_{w,q,i} = 0\,, i \in (V - T) \tag{7}$$

$$d_i = 0\,, i \in (V - D) \tag{8}$$

$$f_{q,i} = 1, q \in Q, i \in T' \tag{9}$$

$$dz_i = 1\,, i \in D' \tag{10}$$

$x_{w,i,j}\,, z_{i,j} \ge 0\,, w \in W, (i,j) \in A$

$y_{w,q,i} \ge 0\,, w \in W, q \in Q, i \in T,$

$d_i \ge 0\,, i \in D,$

$f_{q,i} \in \{0,1\}\,, q \in Q, i \in T, dz_i \in \{0,1\}\,, i \in D$

The cost objective minimizes the total cost of transporting hazardous wastes and waste residues and the fixed annual cost of opening a treatment technology and a disposal facility. The risk objective minimizes the transportation and site risk. Transportation risk is measured with population exposure while site risk is quantified as a function of the amounts of hazardous wastes and waste residues available at those centers and the number of people living within a given radius of these centers. The first constraint is the flow balance constraint for hazardous wastes. This constraint ensures that all generated non-recycled hazardous waste is transported to and treated at a treatment facility. The second constraint is the mass and flow balance constraint for waste residue. The treated and nonrecycled hazardous waste is transformed into waste residue by this constraint, which also ensures that the entire generated and non-recycled waste residue is transported to a disposal site and disposed of. The third and fourth constraints are capacity constraints. That is, the amount of hazardous wastes treated at a treatment technology should not exceed the given capacity of that treatment technology, and the amount of waste residue disposed of in a disposal facility should not exceed the capacity of that disposal facility. The fifth constraint is the minimum amount of requirement constraint. A treatment technology is not established if the minimum amount of waste required for that technology is not exceeded. The sixth constraint is the compatibility constraint, which ensures that a hazardous waste type is treated only with a compatible treatment technology. Constraints (9) and (10) include the existing treatment and disposal facilities.

4   Empirical Approach

The Casablanca-Settat region covers an area of 19,448 km^2 and has 6,862 thousand of inhabitants, a density of 353 inhabitants per km^2 and an area of 2.7% of the national territory. The region has two prefectures Casablanca and Mohammedia and seven provinces: Settat, El Jadida, Ben Slimane, Mediouna, Nouaceur, Berrechid, and Sidi Bennour. The number of communes is 153, of which 29 urban and 124 rural, or about 10% of all municipalities at the national level. The population of the communes ranges from 2645 to 454 908 people. The region has a very satisfactory road structure compared to the rest of the national territory with a linear of 5693 km (Fig. 1).

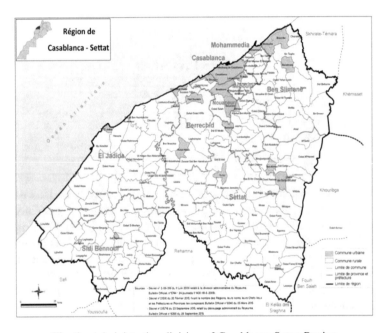

Fig. 1. Administrative division of Casablanca-Settat Region

An ulterior study made by Department of environment, determined seven candidate sites for undesirable facilities: 4 located to the west of Berrechid (B1, B2, B3, B4) and 3 south Mohammedia (M1, M2, M3). Three types of hazardous wastes are considered: compatible with incineration technology, compatible with chemical technology, and compatible with both technologies.

5 Findings

The multi-objective technique adopted is a linear composite objective function combining cost and risk to define the impedance for each link. Cost and risk are scalarized by dividing the terms by their maximums.

$$(p \times Cost/Max\,Cost) + ((1 - p) \times ((Risk/Max\,Risk))$$

Different cases (or alternatives) are given by varying the minimum amount of waste to be processed at a treatment technology to be opened. Case 1 is when the model opens one treatment technology of each type and one disposal facility. The other scenarios are defined altering the number of treatment technology (Table 1).

Table 1. Parameters of the three cases in the application.

Cases	Minimum amount for incinerator (tons)	Expected number of incinerators	Amount for chemical treatment (tons)	number of chemical treatments	Number of disposal centers
1	7000	1	3000	1	1
2	3500	2	1500	2	1
3	3500	2	1500	2	2

Table 2 is the result of our experiment. It represents the optimum locations of treatment technologies and disposal centers for the 7-candidate solution in all of the three cases. If the minimum cost solution ($p = 1$) is adopted, the corresponding risk value is too high, and if the minimum risk solution ($p = 0$) is adopted, the corresponding cost value is again high. The decision maker should choose the parameter between $p = 0.25$ and $p = 0$.

Table 2. Optimum locations of treatment and disposal centers with 7 candidate sites.

Parameter	Treatment technology	Case 1	Case 2	Case 3
P = 0	Incinerator plant	M2	M3, B2	M2, B1
	Chemical treatment	M2	M1, B3	M3, B4
	Disposal center	M1	M2	M2, B2
P = 0.25	Incinerator plant	M1	M3, B4	M2, B1
	Chemical treatment	M2	M2, B4	M1, B4
	Disposal center	M2	M3	M1, B1

(*continued*)

Table 2. (*continued*)

Parameter	Treatment technology	Case 1	Case 2	Case 3
P = 0.5	Incinerator plant	M2	M3, B2	M2, B2
	Chemical treatment	M2	M2, B2	M3, B4
	Disposal center	M3	B1	M3, B1
P = 0.75	Incinerator plant	M1	M3, B2	M2, B1
	Chemical treatment	M3	M1, B3	M1, B1
	Disposal center	M2	M3	M2, B3
P = 1	Incinerator plant	M2	M3, B4	M1, B4
	Chemical treatment	M2	M2, B1	M1, B2
	Disposal center	M2	M1	M2, B4

6 Conclusion

We have presented an empirical validation of Alumur and Kara [2] model with the addition of perceived risk, which allows decision maker to decide where to open treatment centers and with which technologies, where to open disposal centers, how to route different types of hazardous wastes to which of the treatment technologies, and how to route the generated waste residues to disposal centers.

Future research might investigate more risk analysis and incorporate additional objectives and compare results with current work.

As another research direction, one may consider stochastic demand (i.e. amount HW generated at each node), cost and vehicle routing to give more realistic formulation of hazardous waste location routing.

References

1. Alumur, S., Kara, B.Y.: A new model for the hazardous waste location-routing problem. Comput. Oper. Res. **34**(5), 1406–1423 (2007)
2. Diagnostic de l'état de GDND et évaluation du gisement pouvant être valorisés en tant que combustibles secondaires. Report (2013)
3. Giannikos, I.: A multiobjective programming model for locating treatment sites and routing hazardous wastes. Eur. J. Oper. Res. **104**, 333–342 (1998)
4. Jacobs, T.L., Warmerdam, J.M.: Simultaneous routing and siting for hazardous-waste operations. J. Urban Plann. Dev. **120**(3), 115–131 (1994)
5. List, G., Mirchandani, P.: An integrated network/planar multiobjective model for routing and siting for hazardous materials and wastes. Transp. Sci. **25**(2), 146–156 (1991)
6. Rabbani, M., et al.: Using metaheuristic algorithms to solve a multiobjective industrial hazardous waste location-routing problem considering incompatible waste types. Journal of Cleaner Production (2017)
7. Nema, A.K., Gupta, S.K.: Optimization of regional hazardous waste management systems: an improved formulation. Waste Manage. **19**, 441–451 (1999)

8. Samanlioglu, F.: A multi-objective mathematical model for the industrial hazardous waste location-routing problem. Eur. J. Oper. Res. **226**(2), 332–340 (2013)
9. Wyman, M.M., Kuby, M.: Proactive optimization of toxic waste transportation, location and technology. Location Sci. **3**(3), 167–185 (1995)
10. Zhang, Y., Zhao, J.: Modeling and solution of the hazardous waste location routing problem under uncertain conditions. In: ICTE 2011. American Society of Civil Engineers (2011)
11. Zografos, K.G., Samara, S.: Combined location-routing model for hazardous waste transportation and disposal. Transp. Res. Rec. **1245**, 52–59 (1990)

Review of Ontology Based Approaches for Web Service Discovery

Mourad Fariss[✉], Naoufal El Allali, Hakima Asaidi,
and Mohamed Bellouki

MASI Laboratory, FPN B.P 300, Selouane, 62700 Nador, Morocco
m.fariss@ump.ac.ma

Abstract. The Growth in the number of web services leads to rise of problems; therefore, users have a difficulty in finding a web service that has been developed and published. An important issue in web service is the discovery of web services. Semantic technologies facilitate specialization and generalization of service needs as well as service composition. Thus, a higher degree of automation and more precise results can be achieved. In this paper, we are going to provide a review of current ontology approaches for Web Service Discovery.

Keywords: Web service · Web service discovery · Semantics · Ontology · OWL-S

1 Introduction

During the past decade, a lot of researchers have directed significant interest towards web services, an important paradigm of Service–Oriented Architecture (SOA). SOA is an emerging technology in the development of loosely coupled distributed applications on the web. Web services are one of the techniques to implement SOA, a software system designed to support interoperable machine-to-machine interaction over a network (i.e. Browser and Platform independent). Web services convert software applications into web applications, provide loose-coupling at middle-ware level and open an interface to consumers without making them aware of the underlying technologies or implementation details.

The concept of web service based on a series of technology criteria such as SOAP, UDDI and WSDL, is a service-oriented architecture technique. It provides services by standard web protocol to implement the interaction of application services between different platforms. The web service architecture consists of three entities; the service provider, the service registry and the service consumer. The service provider develops or simply offers the web service. The service provider needs to describe the web service in a standard format, which is often XML, and publish it in a central service registry. The service registry contains additional information about the service provider, such as address and contact of the providing company, and technical details about the service. The service consumer retrieves the information from the registry and uses the service description obtained to bind to and invoke the web service (Fariss et al. 2018) (Fig. 1).

F. Khoukhi et al. (Eds.): AIT2S 2018, LNNS 66, pp. 78–87, 2019.
https://doi.org/10.1007/978-3-030-11914-0_8

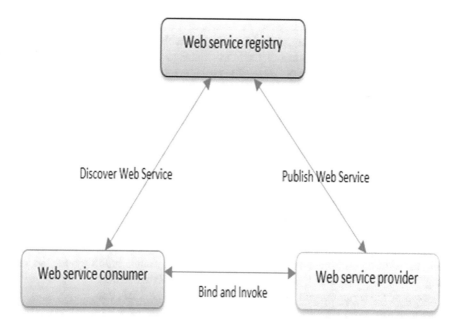

Fig. 1. Web service model

The web service discovery is the user in some way to search the different types of web services and can get its all aspects of specific information. However, the XML based specifications support only the syntactical descriptions of the functionality supported by the services. Therefore, this still requires human interventions for the web service discovery. The evaluation of semantic web has motivated researchers to enrich the web services with the semantic information, which is interpretable by machines. This automates the core activities of web services like discovery, composition, selection and invocation.

This paper is organized in the following sections: Sect. 2 provides a brief overview of the current challenges in web service discovery. Section 3 introduces the semantic web service discovery methods. Various studies on Web Service Discovery using ontologies provided in Sect. 4. And Sect. 5 concludes this paper.

2   Web Service Discovery

The complexity of web services vary in function from simple applications such as weather reports, currency convertors, credit checking, credit card payment, etc. to complex business applications like those of Online Book stores, insurance brokering system, online travel planners etc. Current availability of web services in repositories such as UDDIs, Web portals (e.g. Xmethods, webservicex, webservicelist etc.), over the Internet, and a further scope of rapid future increases makes discovery of web services an important issue from user's perspective.

The discovery of potential web services can be possible mainly through two approaches; one is by using centralized repositories such as UDDI's and another one is by using web crawling techniques. UDDI constitutes metadata about web services and fulfils advertisement requirements to service providers. It also provides search facilities to users for publishing and invoking services. Various standards such as WSDL, SOAP, and UDDI have been developed to support discovery of web services, but pure syntactic nature of all these standards yields in efficient search mechanisms. Since it has been observed that current discovery mechanisms are limited in their search capabilities as they are mostly based on the keywords based matching. The web service consumer searches for a web service in UDDI registry and submits requirements using keywords. Thus, a different mechanism is needed, one that entails locating web services on the basis of the capabilities they provide. Semantics techniques in web service play an important role in seamless integration of diverse services which are based on different terminologies. For example, a service named "car" may not be returned from the query "automobile" submitted by a user, even they are obviously the same at the conceptual level. In addition of semantics to the web services through ontology related technologies can improve the above mentioned syntactic related search.

Some of the other current challenges faced by web services are:

- Considerably web services available on intranet, Internet or private domains are quite high, and they make it essential to have a very accurate process to find the web services needed by the user.
- Retrieval of services with similar functionality as required by a consumer neither much supported as UDDI supports keyword based matching, nor as context based matching.
- There is no support for manual annotation of web services descriptions.
- Lack of intimate knowledge about semantic languages and related toolkits for web services.
- Establishing web service discovery processes.
- There are several existing approaches or tools to allow the web service discovery; however, these approaches often lack different attributes such as quality of service, reusability, incorporation of users' comments or annotations.

All these prevent the user from selecting a web service for their efficient use (Bhardwaj and Sharma 2016).

3   Semantic Web Service Discovery Methods

The term "ontology" is used in many fields such as Artificial Intelligence, Software engineering, Information architecture and many more. Ontology is a structural framework for organizing information or knowledge representations about the world or part of a world, which describes a set of concepts, its properties and relationships among them. Ontology is a "formal specification of a shared conceptualization". That provides a shared vocabulary, and a taxonomy of a particular domain which defines objects, classes, its attributes and their relationships.

Ontology provides semantics to web services and its resources and can greatly improve search mechanisms. In order to realize the vision of semantic web, researchers have proposed several languages, algorithms and architectures. Semantic web services are a component of the semantic web because of markup use which makes data readable by machine (Malaimalavathani and Gowri 2003). Semantic web services use standard such as OWL-S, WSDL-S, WSMO, OWLS-LR and others.

OWL-S: is an ontology language to describe web services; it is the combination of web services and semantic web, mainly to realize using semantics to describe web service. OWL-S including three components: Service Profile, Service Model and Service Grounding. Service Profile describes the service features, Service search agent through the Service Profile to realize Service matching, and provides a superclass of every type of high-level description of the service.

WSDL-S: Current WSDL standard operates at the syntactic level and lacks the semantic to represent the requirements and capabilities of Web Services. WSDL-S is a lightweight approach for adding semantics to Web services. In WSDL-S, the semantic models are maintained outside of WSDL documents and are referenced from the WSDL document via WSDL extensibility elements.

WSMO: WSMO provides a conceptual framework and a formal language to describe all relevant aspects of Web services to facilitate the automation of service discovery using semantics. The overall structure of WSMO is divided into four main elements.

OWL-S is the most dominant approach, though there are several limitations in the OWL-S formalism, e.g. the lack of mediator support, pose challenges in handling real-life situations. Towards addressing this, recent works have explored extensions to the OWL-S framework. Ferndriger et al. (2008) extend OWL-S to define and maintain inheritance relationship between services. Le et al. (2009) propose an ontology comparison strategy to support services based on different OWL-S ontologies (Fig. 2).

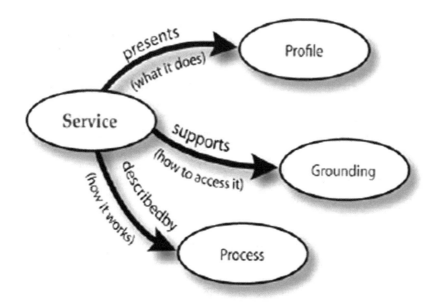

Fig. 2. Service ontology OWL-S (Liu et al. 2010).

4 Ontology Based Approaches for Web Service Discovery

Many of the discovery strategies use the QoS based approaches to refine the search. Pathak et al. (Pathak et al. 2005) presented a framework using ontology-based flexible discovery of semantic web services. They demonstrated how a user's query can be transformed into queries which can be processed by a matchmaking engine.

Adala et al. (2011) proposed a framework that allows semantic web service discovery based on natural language keywords. They employed Natural Language Processing (NLP) to match user query, which is expressed in natural English language with the semantic web service description. They also presented an efficient matching procedure to figure out the semantic distance between ontological concepts.

Paliwal et al. (2012) presented an integrated approach for automated service discovery. Their approach is based on semantic-based service categorization and semantic based service selection. For service categorization, they proposed an ontology based categorization of web services into the functional categories. This lead to better service discovery by matching the service request with an appropriate service description.

Tian (Qiu and Li 2008) proposed an OWL-S based approach for web service discovery with the improved performance (recall and precision) of web service search. Their approach uses service profile description for service matching and discovery. The main WSDL file properties used for matching include service name, inputs, outputs, service parameters and QoS related properties. They presented the integration architecture with semantic matchmaker based on OWL-S mechanism into the UDDI framework. Furthermore, they described web service publishing and discovery using the proposed semantic matchmaker. They also described the matchmaker algorithm.

Klusch et al. (2006) proposed an approach for hybrid semantic web service matching methodology in OWL-S called OWLS-MX. This approach exploits both logic based reasoning and Information Retrieval (IR) methods. Their experimental evaluation concluded that building semantic web service matchmakers solely on description logic reasoners may be insufficient. For future works, they proposed development of more powerful approaches to service matching in the semantic web across disciplines.

In this paper, the authors propose an ontology-based OWL-S extension to adding QoS to web service descriptions (Rohallah et al. 2013). They used an efficient semantic service matching which takes into account concept properties to match concepts in web service and service customer request descriptions. Their approach is based on an architecture composed of four layers: Web service and request description layer, functional match layer, QoS computing layer and reputation computing layer. Future work included to compose the functionality of several web services into one composite and realize consumer preference and QoS-based web service composition.

It has been shown that how the non-functional requirements (QoS) can be incorporated into the service discovery approach so as to generate a list of services including user functional requirements in (Koul et al. 2005). They proposed an ontology-based flexible framework for semantic web services discovery. Their approach is based on user-supplied, context-specific mappings from user ontology to relevant domain ontologies to specify web services. They show how a user's query with certain selection criteria can be transformed into queries processed by a matchmaking engine. They plan to extend proposed approach to service discovery, service invocation and workflow composition for specific data-driven applications.

A logical discovery framework based on semantic description of the capability of web services to improve the speed and accuracy of automated web service discovery has been presented in (Ghayekhloo and Bayram 2015). Their framework improves discovery performance by adding two prefiltering stages to the discovery engine and also tackles the scalability problem. In the first stage, they compared ontology of the user request and Web service categories. In second stage, they reduced number of comparisons as the web services are eliminated based upon a decomposition and analysis of concept. They evaluated the proposed approach using a new web service repository, called WSMO-FL test collection. Logical inference is used for matching, which ensures that the user request is satisfied by the selected web service.

Al Shaban and Haarslev (2005) addressed the problem of service matching requested by users by making use of OWL ontologies and the OWL reasoner RACER. They proposed framework called DECAF which integrates an agent for matchmaking, which employs OWL-S for matching the requirements with web service descriptions. They presented a new way to integrate ontologies with agents and reasoning the ontologies using RACER. They expect that their prototype can be a good start point for switching to the emerging semantic web and using ontologies with mobile agents in a wider application area.

Baklouti et al. (2015) provided a solution to the problems related to the lack of semantic knowledge within the LingWS description. Indeed, they proposed an OWL-S extension to integrate the nonfunctional linguistic properties and relations between them, and they presented a proposal promoting the discovery of the semantic LingWS. Indeed, they extended a semantic matchmaker, called OWLS-MX matchmaker, which has become appropriate to discover LingWS using non-functional linguistic properties (Table 1).

From the previous study, it has been observed that semantic based approaches are simpler and provide better accuracy, but there are some disadvantages as well like less flexibility and more processing time which affects the performance. Table 2 shows the comparison between ontology and other techniques on the performance parameters.

Challenges of ontologies in web service discovery:

- There has been less research work in the field of semantic web service that targets to qualitatively enhance or upgrade web service ontologies or can facilitate the use of ontologies and improve the selection process.
- It is a cumbersome task to develop a common ontology for countless web services.
- It has been observed that there are almost no information regarding the challenges the researchers or projects faced while using the ontologies.
- Furthermore, the majority of already existing services does not support the associated semantics and there are challenges on the potential conversion of existing non-semantic structure to the semantic descriptions.

Table 1. Comparison of ontology based approaches

Approach	Ontology supported	Framework/Approach name	Scalable/Flexible	QOS used	Specific Tool used/Middleware are/Agent used/Reasoning	Ranking used	Annotation support	Ontology Mapping used	Request expansion	NLP Used
(Pathak et al. 2005)	OWL-S	NS	Yes	Yes	JESS	Yes	No	Yes	No	No
(Adala et al. 2011)	OWL-S/WSMO	NS	Yes	No	Wordnet/SUMO	Yes	No	Yes	No	Yes
(Paliwal et al. 2012)	SUMO	NS	NS	No	Cosine similarity	Yes	No	Yes	Yes	Yes
(Qiu and Li 2008)	OWL-S	UDDI	NS	Yes	NS	No	No	Yes	No	No
(Klusch et al. 2006)	WSMO	NS	No	Yes	NS	NS	Yes	No	No	No
(Rohallah et al. 2013)	OWL-S	NS	No	Yes	q-grams	Yes	Yes	No	No	No
(Al Shaban and Haarslev 2005)	OWL-S	DECAF	Yes	No	RACER	No	No	No	No	No
(Ghayekhloo and Bayram 2015)	WSMO	NS	Yes	No	Logical Inference	Yes	No	No	No	No
(Koul et al. 2005)	OWL-S	NS	NS	Yes	Jess & Jena	Yes	No	Yes	No	No
(Baklouti et al. 2015)	OWL-S	LingWS	NS	No	NS	Ns	No	Yes	No	No

*NS- Not Specified

Table 2. Comparison of semantic based and syntactic based approaches

Features	Semantic approach	Syntactic/other approach
Accuracy	High	Low
Complexity	High	Less
Flexibility	High	Low
Processing time	More	Less
Reliability	High	Less

- Semantic web services have not yet been fully adopted by the industry. One of the major challenges is the unavailability of semantically annotated content for use.
- Since existing web services follow mostly WSDL specifications, they should be upgraded to semantic web services. This can be achieved with the ability to automatically convert WSDLs into their semantic counterparts like SAWSDL or OWL-S.

5   Conclusion

Discovery system plays an important role in web service model for efficient web service retrieval. As web service discovery involves manual intervention, it is a cumbersome and time consuming task to find the web services as needful as the consumer expects. This necessitates the resolutions for automatic discovery techniques and takes attention of researchers. Currently, OWL-S based ontology language is the most standardized and perhaps is most comprehensive semantic web service technology deployed. This paper presents a survey of the state–of-the-art on semantic web service discovery systems. We reviewed more recent discovery works outlining the main approaches and discussing representative systems for each approach. Though our survey does not consider all published works.

It is observed that following areas can be further worked upon as highlighted below:

- Better matchmaking mechanisms.
- Use of NLP techniques for better feature selections.
- User query expansion techniques.
- Improving and better use of existing ontologies standards.
- Use of hybrid approaches for performance and accuracy improvement.

This study will help researchers to select the best and appropriate ontology-based approach for web service discovery, and they can propose a new approach to reduce their disadvantages. For us, we focus on exploiting this study for a new system discovery and selection web services.

References

Adala, A., Tabbane, N., Tabbane, S.: A framework for automatic web service discovery based on semantics and NLP techniques. Adv. Multimed., 1–7 (2011). Hindawi Publishing Corp., https://doi.org/10.1155/2011/238683

Al Shaban, A., Haarslev, V.: Applying Semantic Web Technologies to Matchmaking and Web Service Descriptions. http://scholar.googleusercontent.com/scholar?q=cache:C-0VaUb_FQ0J:scholar.google.com/+A.+A.+Shaban+and+V.+Haarslev,+Applying+Semantic+Web+Technologies+to+Matchmaking+and+Web+Service+Descriptions",+&hl=fr&as_sdt=0,5. Accessed 30 Sep 2018

Baklouti, N., Gargouri, B., Jmaiel, M.: Semantic-based approach to improve the description and the discovery of linguistic web services. Eng. Appl. Artif. Intell. **46**, 154–165 (2015). Pergamon, https://doi.org/10.1016/J.ENGAPPAI.2015.09.005

Bhardwaj, K.C., Sharma, R.K.: Ontologies: a review of web service discovery techniques. Int. J. Energy Inf. Commun. **7**(5), 1–12 (2016)

Fariss, M., Asaidi, H., Bellouki, M.: ScienceDirect comparative study of skyline algorithms for selecting web services based on QoS. Procedia Comput. Sci. **127**, 408–415 (2018). https://doi.org/10.1016/j.procs.2018.01.138

Ferndriger, S., Bernstein, A., Dong, J.S., Feng, Y., Li, Y.-F., Hunter, J.: Enhancing semantic web services with inheritance. In: pp. 162–177. Springer, Heidelberg (2008). https://doi.org/10.1007/978-3-540-88564-1_11

Ghayekhloo, S., Bayram, Z.: Prefiltering strategy to improve performance of semantic web service discovery. Scientific Programming 2015. Hindawi Publishing Corp., pp. 1–15 (2015). https://doi.org/10.1155/2015/576463

Klusch, M., Fries, B., Sycara, K.: Automated semantic web service discovery with OWLS-MX. In: Proceedings of the Fifth International Joint Conference on Autonomous Agents and Multiagent Systems - AAMAS 2006, vol. 915. ACM Press, New York (2006). https://doi.org/10.1145/1160633.1160796

Koul, N., Doina, C., Honavar, V.: Discovering Web Services over the Semantic Web. Computer Science Technical Reports Computer Science, vol. 237 (2005). http://lib.dr.iastate.edu/cs_techreports/237

Le, D.N., Tran, B.D., Tan, P.S., Goh, A.E.S., Lee, E.W.: MODiCo: a multi-ontology web service discovery and composition system. In: pp. 531–534. Springer, Heidelberg (2009). https://doi.org/10.1007/978-3-642-02818-2_55

Liu, F., Shi, Y., Yu, J., Wang, T., Wu, J.: Measuring similarity of web services based on (2010). https://doi.org/10.1109/ICWS.2010.67

Malaimalavathani, M., Gowri, R.: A survey on semantic web service discovery. In: International Conference on Information Communication and Embedded Systems, pp. 222–225 (2003)

Paliwal, A.V., Shafiq, B., Vaidya, J., Xiong, H., Adam, N.: Semantics-based automated service discovery. IEEE Trans. Serv. Comput. **5**(2), 260–275 (2012). https://doi.org/10.1109/TSC.2011.19

Pathak, J., Koul, N., Caragea, D., Honavar, V.G.: A framework for semantic web services discovery. In: Proceedings of the Seventh ACM International Workshop on Web Information and Data Management - WIDM 2005, vol. 45. ACM Press, New York (2005). https://doi.org/10.1145/1097047.1097057

Qiu, T., Li, P.: Web service discovery based on semantic matchmaking with UDDI. In: 2008 The 9th International Conference for Young Computer Scientists, pp. 1229–1234. IEEE (2008). https://doi.org/10.1109/ICYCS.2008.474

Rohallah, B., Ramdane, M., Zaidi, S.: Agents and Owl-s based semantic web service discovery with user preference support, June 2013. https://doi.org/10.5121/ijwest.2013.4206

An Ontological Representation of ITIL Framework Service Level Management Process

Abir El Yamami[✉], Khalifa Mansouri, Mohammed Qbadou,
and Elhossein Illoussamen

Laboratory: Signals, Distributed Systems and Artificial Intelligence (SSDIA),
ENSET Mohammedia, University Hassan II Casablanca, Casablanca, Morocco
Abir.elyamami@gmail.com, khmansouri@hotmail.com,
qbmedn7@gmail.com, illous@hotmail.com

Abstract. IT Service Management (ITSM) is an information system management approach that represents the information system in the form of a set of capabilities that bring value to customers in the form of services. ITIL (Information Technology Infrastructure Library) framework has positioned itself as a generic solution to tackle a broad range of ITSM issues and try to guide IT managers in their endeavors. Research on the subject has been mostly restricted to the implementation of the incident, problem and change management processes. Such approaches, however, have failed to address the deployment of the service level management (SLM) process. Yet, the SLM process has been criticized as the most important process in ITIL framework and that is highly dependent on the other processes. In this paper, an ontological metamodel of ITIL SLM process explicating its core concepts is presented; the goal is to provide a machine-readable document for modeling SLM domain knowledge according to ITIL V3 Framework. The proposed metamodel could be used for: (i) a deeper evaluation for comparing it to other ITSM models, and (ii) the integration of different frameworks and standards of IT governance such as Cobit and ISO 27001.

Keywords: IT Service Management · IT governance · ITIL · Ontology · Protégé

1 Introduction

ITSM can be defined as a set of processes that cooperate to ensure the quality of IT services according to the levels of services agreed by the customers [1]. ITIL is the most popular framework dedicated to ITSM [2], it combines a set of good practices for managing IT information, collected from a variety of sources around the world. It was developed in the UK in the 1980s by the government to improve the management of its IT. ITIL covers a broad field of governance of the information system by focusing on the concept of service and quality. It uses the concept of service contract between service seekers and service providers [3].

© Springer Nature Switzerland AG 2019
F. Khoukhi et al. (Eds.): AIT2S 2018, LNNS 66, pp. 88–94, 2019.
https://doi.org/10.1007/978-3-030-11914-0_9

This paper is an effort to create a common representation of ITSM good practices based on the collection and analysis of data contained in the official guides of ITSM frameworks. This representation should be formal, complete and reusable in order to be used by intelligent applications, other knowledge domains or IT experts. It proposes an ontological representation of SLM process domain knowledge based on ITIL V3 framework. We describe below the design process followed for the development of our proposed ontology:

- **Motivation:** IT service management applications are rarely developed taking into consideration the guidelines and best practices specified in information system frameworks. As a result, this work attempts to deliver to developers a common ontology of ITIL SLM practices.
- **Objective:** The objective is to provide a machine-readable document for modeling IT SLM domain.
- **Design and development:** Proposing an ontological approach to the adoption of ITIL practices to ensure that IT services are continuously maintained.
- **Demonstration:** The proposed artefact was evaluated for its inconsistency, incompleteness, and redundancy. The ontology of SLM domain has been implemented in protégé software, and the integrity has been validated by the inference engines: Fact++ 1.6.5 and Pellet.

The rest of this paper is organized as follow: Sect. 2 depicts our proposed artefact, Sect. 3 presents the evaluation of the proposed meta-model, and finally the results of our contribution are presented in Sect. 4.

2 Methodology

The service level management (SLM) is the process of documenting and agreeing service targets in Service Level Agreements (SLA), then monitoring and reviewing the actual service levels against those targets.

An ontology is a computational model of some portions of the world. It is a collection of key concepts and their inter-relationships, collectively giving an abstract view of an application domain [4]. It aims at interweaving human understanding of symbols with their machine process ability.

The relevance of ontologies has been recognized in several practical fields, such as natural language translation, agent systems, geographic information systems, biology and medicine, knowledge management systems and e-commerce platforms. In this sense, [5] presented an automatic ontology-based knowledge extraction from Web documents to create personalized biographies. [6] proposed an ontology-based knowledge management system to assist engineers in sharing and managing knowledge on water quality modelling. [7] presented an ontology-based search engine for the semantic web. [8] developed an ontology-based intelligent web portal system for recruitment tasks. They utilized an ontology to represent the knowledge of the recruitment domain. [9] presented a dynamic invocation of semantic web services using unfamiliar ontologies.

Likewise, an ontological meta-model of ITIL framework SLM process will be presented. The motivation behind using an ontology for SLM is to:

- **Share the common understanding of the information structure between people or software manufacturers** [10]: ITSM applications must share the same ontology, based on best management practices for ITSM.
- **To allow the reuse of knowledge on a domain:** it would be possible to integrate several existing ontologies to build a global ontology of IT service governance domain.
- **Explain what is considered implicit on a domain:** the explanation of the postulates of ITSM domain makes it possible to modify these specifications in case of evolution of the domain knowledge.

In order to develop our ontology, a collection, organization, analysis and presentation of the data was made. The official documentation of the ITIL V3 guides has been used as sources of information. We present below the ontological description this domain knowledge.

2.1 Identification of Service Requirements

Process Objective: To capture requirements from the consumer viewpoint for new IT services or IT service level modification. The objective is to improve relationships with satisfied consumers so that IT effort can be focused on those areas that the consumer thinks are key.

Process Description: A service is defined by a service name and service description, there is 3 types of services: Application services, technical services and professional services [11]. For each service, 3 service levels should be defined: Gold, Silver and Bronze. The service levels are specified according to quality of service metrics. In this phase, ITIL points out that the consumer can initiates a request for a service level modification. To do so, the consumer should define the QoS requirements such as availability requirements, performance requirements, continuity, and service desk (Fig. 1).

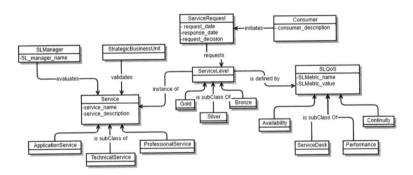

Fig. 1. Management of service level requirements process ontology

2.2   Agreements Sign-off and Service Activation

Process Objective: To check service acceptance and make sure that the service contracts are signed off by both provider and consumer.

Process Description: After the service level metrics are fulfilled, ITIL points out that a contract between the service consumer and service provider is used as a basis for charging or cost allocation and helps demonstrate what value consumers are receiving for their money. In particular, this process makes sure that all relevant OLAs (operational level agreement) are signed off by their Service Owners, and that the SLA (service level agreement) is signed off by the consumer. Figure 2 depicts the ontological representation of the SLM Process: "*Agreements Sign-Off and Service Activation*" domain knowledge.

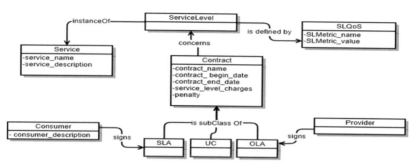

Fig. 2. Agreement sign off and service activation process ontology

2.3   Service Level Monitoring and Reporting

Process Objective: To monitor achieved service levels and compare them with agreed service level targets in order to improve service quality.

Process Description: In this phase, the service level manager initiates a service monitoring session, the objective is to identify root cause for weak areas so that remedial action can be taken (root cause resolutions), thus improving future service quality. The service level manager should send satisfaction surveys to the consumer frequently in order to identify where working efficiency can be improved. After the service level monitoring session, a service level report is generated and is circulated to consumer and all other relevant parties, as a basis for measures to improve service quality. Figure 3 depicts the ontological representation of the SLM Process: "*Service Level Monitoring and Reporting*" domain knowledge.

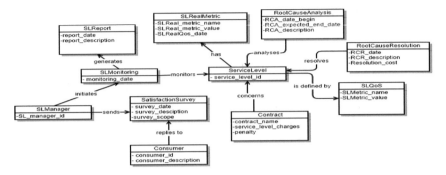

Fig. 3. Service level monitoring and reporting process ontology

3   Artefact Evaluation

In order to evaluate our proposed artefact for its inconsistency, incompleteness, and redundancy, ITIL SLM ontology has been implemented in protégé software [12], and the integrity has been validated trough the inference engines: Fact++ 1.6.5 [13] and Pellet [14] integrated as plug-in in protégé software. Figure 4 represents the generated inferred classes.

Fig. 4. Protégé generated classes

4 Conclusion

Since a complete ontology for ITSM does not exist in the literature, it seems necessary to construct a shared representation of the manipulated concepts. The objective is to contribute to the professional literature by providing a machine-readable document for the IT governance knowledge area, and secondly to the scientific literature interested in improving IT governance frameworks and standards by increasing the understanding of the architectures of these repositories.

As presented in this paper, we used the ITIL framework to design a SLM ontology for IT services. Yet the adoption of ITIL knowledge domains is not an easy task as their definition and role is unclear. This contribution presented an alternative for an ITSM system, it provides an ontology, aimed at formalizing the knowledge of this field. Protégé software was used in the design and evaluation of the proposed ontology. The aim is to eliminate misunderstanding of the objective behind the implementation of ITSM models.

By sharing ITIL SLM knowledge, ITSM expertise is leveraged across the organization and the communication between different stakeholders is enhanced. The results should aid in the development of IT governance approaches for use by the non-specialist IT governance frameworks in organizations, and to limit IT frameworks bureaucracy as far as possible.

As a perspective, this ontology can be useful for the integration of IT Governance frameworks: COBIT and ITIL practices are closely aligned and can be used together for the governance of information systems.

References

1. Young, C.M.: An introduction to IT service management (2004)
2. El Yamami, A.: Toward a new multi-agents architecture for the adoption of ITIL framework by small and medium-sized enterprises. In: 4th IEEE International Colloquium on Information Science and Technology (CiSt), Tangier (2016)
3. El Yamami, A., Ahriz, S., Mansouri, K., Qbadou, M., Illoussamen, E.: Developing an assessment tool of ITIL implementation in small scale environments. Int. J. Adv. Comput. Sci. Appl. (IJACSA) **8**(9), 183–190 (2017)
4. Lee, C.S., Jian, Z.W., Huang, L.K.: A fuzzy ontology and its application to news summarization. IEEE Trans. Syst. Man Cybern. Part B (Cybern.) **35**(5), 859–880 (2005)
5. Alani, H., Kim, S., Millard, D., Weal, M.J., Hall, W., Lewis, P.H., Shadbolt, N.R.: Automatic ontology-based knowledge extraction from web documents. IEEE Intell. Syst. **18**(1), 14–21 (2003)
6. Chau, K.W.: An ontology-based knowledge management system for flow and water quality modeling. Adv. Eng. Softw. **38**(3), 172–181 (2007)
7. Corby, O., Dieng-kuntz, R., Gandon, F.: Faron-Zucker, C: Searching the semantic web: approximate query processing based on ontologies. IEEE Intell. Syst. **21**(1), 20–27 (2006)
8. García-Sánchez, F., Martínez-Béjar, R., Contreras, L., Fernández-Breis, J.T., Castellanos-Nieves, D.: An ontology-based intelligent system for recruitment. Expert Syst. Appl. **31**(2), 248–263 (2006)

9. Burstein, M.H.: Dynamic invocation of semantic web services that use unfamiliar ontologies. IEEE Intell. Syst. **19**(4), 67–73 (2004)
10. Musen, M.: Dimensions of knowledge sharing and reuse. Comput. Biomed. Res. **25**, 435–467 (1992)
11. El Yamami, A., Mansouri, K., Qbadou, M., Illoussamen, E.: Introducing ITIL framework in small enterprises: tailoring ITSM practices to the size of company. Int. J. Inf. Technol. Syst. Approach (IJITSA) **12**(1), 1–9 (2019)
12. Musen, M.: The Protégé project: a look back and a look forward. AI Matters **1**(4) (2015). https://doi.org/10.1145/2557001.25757003. Association of Computing Machinery Specific Interest Group in Artificial Intelligence, Intelligence
13. Building, K.: "Fact++," School of Computer Science University of Manchester (2017). http://owl.man.ac.uk/factplusplus/
14. Pellet: Clark & Parsia, LLC (2011). http://pellet.owldl.com/

Multimedia Systems and Information Processing

Robust Watermarking for Medical Image Against Geometric and Compression Attacks

Imane Assini$^{(\boxtimes)}$, Abdelmajid Badri, Khadija Safi, Aicha Sahel, and Abdennaceur Baghdad

EEA and TI Laboratory Faculty of Sciences and Techniques (FSTM), Hassan II University of Casablanca, BP 146, Mohammedia 20650, Morocco assini.media@gmail.com, abdelmajid_badri@yahoo.fr, k_saly2000@yahoo.fr, sahel_ai@yahoo.fr, nasser_baghdad@yahoo.fr

Abstract. The security of medical images becomes increasingly important for many reasons such as confidentiality, authentication, and integrity. In this context, the digital watermarking has appeared as a solution to protect the medical data diffused in hospital networks. In this article, we present a robust technique of medical image watermarking of various modalities such as Computed Tomography (CT Scanner), Magnetic Resonance Angiography (MRA), Magnetic Resonance Imaging (MRI), and X-Ray imaging, based on the combination of the Stationary Wavelet Transform (SWT) and the Fast Walsh-Hadamard Transform (FWHT). In our contribution, we have inserted the watermark of size 512×512 in the high-frequency subbands of the original image of the same size in order to increase the protection. The experimental results show that our contribution gives a good robustness against geometric and compression attacks compared to other techniques cited in the literature.

Keywords: Digital watermarking · Medical image · Security · SWT · FWHT

1 Introduction

The digital watermarking is a recent discipline in the medical field, it allows inserting data of the patient into a cover medical image in such a way that it is invisible [1].

To obtain a perfect system of watermarking, we must respect the three following properties: the invisibility of watermark, the capacity and the robustness against several attacks.

The techniques of digital watermarking can be grouped according to the domain of insertion: the spatial domain [2] and the frequency domain [3–5]. Currently, the insertion in the frequency domain can ensure a high level of security of medical data by preserving their integrity and their confidentiality.

Our approach is within this framework and consists in conceiving a system of marking which aims to improve protection of the medical images in terms of robustness, invisibility, and capacity.

© Springer Nature Switzerland AG 2019
F. Khoukhi et al. (Eds.): AIT2S 2018, LNNS 66, pp. 97–103, 2019.
https://doi.org/10.1007/978-3-030-11914-0_10

This article is based on the following sections: Sect. 2 presents the techniques of watermarking used in our method SWT and FWHT. In Sect. 3, we have proposed an algorithm of watermarking based on SWT-FWHT. Section 4, presents the experimental results and finally, a conclusion is drawn.

2 Backgrounds

The proposed technique represents a hybrid medical image watermarking based on the SWT-FWHT combination (Fig. 1).

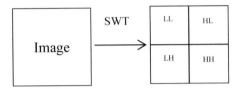

Fig. 1. First level decomposition SWT

The Stationary Wavelet Transform (SWT) also called the undecimated Wavelet Transform [6], has the advantage to provide a good quality of watermarked image and to improve the robustness of the system against several attacks.

The Fast Walsh-Hadamard Transform FWHT [7] offers advantages such as the speediness of the calculation of coefficients of Hadamard and the ease of implementation. The choice of SWT and FWHT it's based on their advantages, so their combination allows developing a new watermarking technique very effective in terms of robustness, capacity and invisibility.

2.1 Stationary Wavelet Transform (SWT)

The Stationary Wavelet Transform (SWT) [6] is similar to the Discrete Wavelet Transform (DWT), except that the signal is not decimated, and there is no under-sampling phase for decomposition and Over-sampling phase for reconstruction. The SWT decomposes an image into four subbands, approximation sub-band LL, and three detail sub-bands: LH, HH, and HL respectively corresponding to vertical, diagonal and horizontal details. In our case, the watermark will be inserted in the HH sub-band.

2.2 Fast Walsh-Hadamard Transform (FWHT)

The Walsh-Hadamard transform [8] is a non-sinusoidal orthogonal transformation which divides a signal into a set of waveforms of values 1 or −1 which are rectangular or square waves called Walsh functions. The Hadamard H_N transform is a matrix of size $2^N \times 2^N$, $N > 0$ determinate by (1):

$$H_N = \frac{1}{\sqrt{2}} \begin{pmatrix} H_{N-1} & H_{N-1} \\ H_{N-1} & -H_{N-1} \end{pmatrix} \qquad (1)$$

Some examples of Hadamard matrices 1×1, 2×2 and 4×4 are defined by (2), (3) and (4) respectively:

$$H_0 = 1 \tag{2}$$

$$H_1 = \frac{1}{\sqrt{2}} \begin{pmatrix} 1 & 1 \\ 1 & -1 \end{pmatrix} \tag{3}$$

$$H_2 = \frac{1}{2} \begin{pmatrix} 1 & 1 & 1 & 1 \\ 1 & -1 & 1 & -1 \\ 1 & 1 & -1 & -1 \\ 1 & -1 & -1 & 1 \end{pmatrix} \tag{4}$$

3 Proposed Technique

The purpose of this paper is to describe and to analyze a new hybrid technique based on the combination of two techniques SWT and FWHT, of which the ultimate goal is to augment the security of medical images. The Sects. 3.1 and 3.2 show the algorithm of insertion and extraction of watermarking.

3.1 Watermark Inserting Algorithm

1. Application of SWT on the cover medical image and the watermark image using the wavelet of Daubechies until the 1st level to get (LL_1, LH_1, HL_1, HH_1) and (LL_{w1}, LH_{w1}, HL_{w1}, HH_{w1}) respectively.
2. Application of FWHT on the selected high frequency subbands HH_1 and HH_{w1}.
3. Insertion of watermark by Eq. (5):

$$\boldsymbol{FWHT(HH_{IW})} = FWHT(HH_1) + \alpha \times FWHT(HH_{W1}) \tag{5}$$

Where α: Scaling factor
4. Calculation of inverse FWHT and inverse SWT to get the watermarked image.

3.2 Watermark Extracting Algorithm

1. Apply the first level of decomposition SWT on the original and the wa-termarked image and then select HH_1 and HH_{IW} sub-bands respectively.
2. Application of FWHT on the selected sub-bands.
3. Extraction of the watermark by Eq. (6):

$$\boldsymbol{W} = (FWHT(HH_{IW}) - FWHT(HH_1))/\alpha \tag{6}$$

4. Calculation of inverse FWHT and inverse SWT to get the extracted watermark.

4 Experimental Results

The proposed SWT-FWHT approach has been implemented in Matlab R2013b. Several tests were applied on the original images of diverse modalities such as Computed Tomography (CT Scanner), Magnetic Resonance Angiography (MRA), Magnetic Resonance Imaging (MRI), and X-Ray imaging of sizes 512 × 512 and the watermark considered as patient image of the same size of the cover medical image shown in Fig. 2.

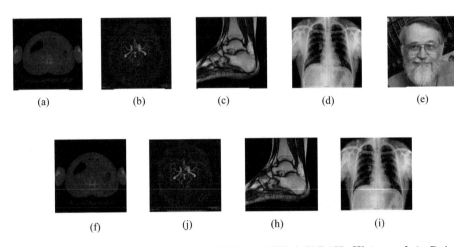

(a)   (b)   (c)   (d)   (e)

(f)   (j)   (h)   (i)

Fig. 2. Original images (a: CT scanner, b: MRA, c: MRI, d: X-RAY), **Watermark** (e: Patient image), **Watermarked images**: (f, j, h, i)

To evaluate the effectiveness of the proposed combination several parameters are taken into consideration such as.

The Mean Squared Error (MSE), the Signal-to-Noise Ratio (PSNR) and the Normalized Correlation coefficient (NC). The PSNR is described as follows [9]:

$$PSNR = 10log\left[\frac{(255)^2}{MSE}\right] \tag{7}$$

Where MSE [9] presents the mean squared error, its measures the quality evaluation between an original image I and a resulting image Î after inserting a watermark image:

$$MSE = \frac{1}{MN}\sum_{i=1}^{M}\sum_{j=1}^{N}\left[I(i,j) - \hat{I}(i,j)\right]^2 \tag{8}$$

I and Î are the cover medical image and the watermarked image of size MxN respectively. Furthermore, the robustness of the algorithm of watermarking is measured by the normalized correlation coefficient (NC) [9]:

$$NC = \frac{\sum_i \sum_j (W(i,j) * W'(i,j))}{\sqrt{(\sum_i \sum_j (W_{(i,j)})^2)(\sum_i \sum_j \left(W'_{(i,j)}\right)^2)}}$$

(9)

W (i, j) and W' (i, j) are the original and the extracted watermark respectively.

The imperceptibility of our system of watermarking is calculated by PSNR. Figure 2 shows the Cover medical images, the watermark image (image of the patient) and the watermarked images. We conclude that our proposed method gives perfect invisibility.

In Table 1 the effectiveness of our contribution has been studied without applying attacks. The maximum PSNR value obtained with the Abdomen CT Scanner image is 47.75 at scaling factor 1. However, the NC correlation values are 1 in all gain factor.

Table 1. The PSNR and NC values for the original images at different Scaling factor

Scaling factor	Abdomen CT scanner		MRA brain		Ankle MRI		Chest RX	
	PSNR	NC	PSNR	NC	PSNR	NC	PSNR	NC
1	47.75	1	43.90	1	47.51	1	42.07	1
5	43.51	1	40.65	1	43.39	1	39.80	1
9	40.60	1	38.02	1	40.45	1	37.59	1

The robustness of our watermarking algorithm is verified by the application of different attacks such as geometric and compression attacks.

In Table 2, the NC performance of our approach against various attacks is compared against existing methods [6, 10, 11]. The maximum NC correlation values at the scaling factor equal to 9 are 1 against Rotation (10, 20, 30, and 50), and compression JPEG (40, 50, and 60).

Table 2. Compression NC performance at gain factor = 9

Attacks	Proposed method	[6]	[10]	[11]
Rotation 10	1	0.9572	–	–
Rotation 20	1	0.9420	–	–
Rotation 30	1	0.9436	–	–
Rotation 50	1	–	0.9988	–
Resize	0.9999	–	0.9987	–
Translation	0.9999	–	–	–
JPEG QF = 40	1	–	–	0.9875
JPEG QF = 50	1	–	0.9990	0.9942
JPEG QF = 60	1	–	–	1

The results obtained show that our approach gives a good compromise invisibility, capacity, and robustness. In case of capacity, the size of watermarks inserted into [6, 10, 11] are 64×64, 256×256 and 32×32 respectively. However, in our approach we have inserted a watermark of size 512×512.

In case of invisibility, the PSNR values obtained by our technique are 47.75 dB and 43.51 dB against 33.7641 dB and 33.9209 dB obtained by [10] at scaling factor 1 and 5 respectively. And finally, the robustness of the proposed method against all attacks shown in Table 2 is important against [6, 10, 11].

5   Conclusion

This article presents a hybrid watermarking technique based on SWT-FWHT for the purpose of ensuring the security of medical images, where the watermark image is inserted into the high frequency HH subbands of the original image. Furthermore, these regions are able to increase the robustness of the watermarking system.

The proposed watermarking technique combines the advantages of SWT and FWHT that offer good results in terms of invisibility, robustness and capacity. Several tests have highlighted our proposed algorithm, it has the advantage to insert an important quantity of data without altering the quality of the image, as well as its robustness against geometric and compression attacks compared to other techniques.

In the future works we will focus on developing the performance of our approach against other types of geometric attacks.

References

1. Assini, I., Badri, A., Safi, K.: Adaptation of different techniques on digital image watermarking in medical domain: a review. Int. J. Adv. Res. Comput. Sci. Software Eng. (IJARCSSE) **5**(12), 5–9 (2015). ISSN: 2277 128X
2. Swanson, Mitchell D., Kobayashi, Mei, Tewfik, Ahmed H.: Multimedia data-embedding and watermarking technologies. Proc. IEEE **86**(6), 1064–1087 (1998). https://doi.org/10.1109/5. 687830
3. Assini, I., Badri, A., Safi, K., Sahel, A., Baghdad, A.: Hybrid multiple watermarking technique for securing medical image using DWT-FWHT-SVD. In: 2017 International Conference on Advanced Technologies for Signal and Image Processing (ATSIP), pp. 1–6. IEEE (2017). https://doi.org/10.1109/atsip.2017.8075569
4. Assini, I., Badri, A., Safi, K., Sahel, A., Baghdad, A.: Robust multiple watermarking technique for medical applications using DWT, DCT and SVD. Trans. Mach. Learn. Artif. Intell. **5**(4) (2017). http://dx.doi.org/10.14738/tmlai.54.3219
5. Scholar, P.G.: A survey: digital image watermarking techniques. Int. J. Sig. Process. Image Process Pattern Recogn. **7**(6), 111–124 (2014)
6. Shokrollahi, Z., Yazdi, M.: A robust blind watermarking scheme based on stationary wavelet transform. J. Inf. Hiding Multimedia Sig. Process. **8**(3), 676–687 (2017)
7. Meenakshi, K., Rao, S., Prasad, K.S.: A Hybridized robust watermarking scheme based on fast walsh-hadamard transform and singular value decomposition using genetic algorithm. International J. Comput. Appl. **108**(11) (2014). https://doi.org/10.5120/18952-9952

8. Gunjal, B.L., Mali, S.N.: Comparative performance analysis of digital image watermarking scheme in DWT and DWT-FWHT-SVD domains. In: India Conference (INDICON), 2014 Annual IEEE, pp. 1–6. IEEE (2014). https://doi.org/10.1109/indicon.2014.7030391

9. Singh, A.K., Dave, M., Mohan, A.: Hybrid technique for robust and imperceptible multiple watermarking using medical images. Multimedia Tools Appl. **75**(14), 8381–8401 (2016). https://doi.org/10.1007/s11042-015-2754-7

10. Singh, S., Singh, R., Singh, A.K., Siddiqui, T.J.: SVD-DCT based medical image watermarking in NSCT domain. In: Quantum Computing: An Environment for Intelligent Large Scale Real Application, pp. 467–488. Springer, Cham (2018). https://doi.org/10.1007/978-3-319-63639-9_20

11. Ernawan, F., Kabir, M.N.: A blind watermarking technique using redundant wavelet transform for copyright protection. In: 2018 IEEE 14th International Colloquium on Signal Processing & Its Applications (CSPA), pp. 221–226. IEEE (2018). https://doi.org/10.1109/cspa.2018.8368716

R Peak Detection Based on Wavelet Transform and Nonlinear Energy Operator

Lahcen El Bouny$^{(\boxtimes)}$, Mohammed Khalil, and Abdellah Adib

LIM@II-FSTM, B.P. 146, 20650 Mohammedia, Morocco
lahcenbouny@gmail.com, medkhalil87@gmail.com, adib@fstm.ac.ma

Abstract. The Electrocardiography is a graphical representation of the heart's electrical activity. In this paper, we present an efficient algorithm for QRS complex detection in the Electrocardiogram (ECG) signal. The proposed algorithm is based on a combination of the Shift Invariant Wavelet Transform (ShIWT), a Nonlinear transform called Nonlinear Energy Operator (NEO) and a simple thresholding function followed by some decision rules for accurate R peak detection. In our scheme, ShIWT was used to filter out the raw ECG signal and the NEO was applied to highlight the QRS complex patterns. Finally, after simple thresholding stage, R peak time positions on the filtered ECG signal can be detected accurately with the help of efficient decision rules. The experimental results and tests carried over real ECG signals taken from the MIT-BIH Arrhythmia Database (MITDB) show that our proposed approach gives a comparable or higher detection performances against the state of the art techniques with an average Sensitivity (Se) of 99.76%, average Positive Predictivity (P+) of 99.77% and a Detection Error Rate (DER) of 0.47%.

Keywords: ECG · QRS complex detection · ShIWT · TEO · NEO

1 Introduction

Electrocardiogram (ECG) signal reflects the different steps of the heart's electrical activity. It is widely materialized by five principal deflections P, QRS and T (and U wave in some cases). Until now, ECG signal has been considered as an important diagnostic tool for cardiologists to detect and to prevent the cardiovascular diseases. In fact, heart diseases are considered as one of the principal cause of death in the world with about 17.3 million death according to the World Health Organization. For this reason, accurate heart disease diagnosis with ECG tool is a challenging task. In this context, QRS complex detection is a crucial step for all kinds of ECG signal processing applications.

Various approaches have been proposed to achieve more accurate QRS complex detection. In general scheme, QRS complex detection algorithms include

© Springer Nature Switzerland AG 2019
F. Khoukhi et al. (Eds.): AIT2S 2018, LNNS 66, pp. 104–112, 2019.
https://doi.org/10.1007/978-3-030-11914-0_11

two important steps [1]. The first step is the pre-processing stage in order to denoise the raw ECG signal affected by different noises and artifacts as well as to highlight the QRS complex and to attenuate the other waves. At the second stage, sophistical decision rules are applied to define the exact R time positions.

Firstly, a large number of approaches based on derivatives function have been widely studied and discussed such as the first algorithm developed by Pan and Tompkins [2]. More sophistical techniques such as neural networks [3] are also established to give better detection performances. Hilbert Transform has also been used in [4]. In the last decades, Time-Frequency approaches including Discrete Wavelet Transform (DWT) has also been employed in various works [5]. The basic idea behind using wavelet transform for QRS detection issue is to find the frequency sub-bands that represent the important information about the QRS complex and those dominated by the noise components or other unlike peaks (e.g. P and T waves). Further, post-processing steps are applied to determine the time positions corresponding to the R peaks.

In this work, an efficient simple tool for QRS complex detection problem was proposed. It is based on Shift Invariant Wavelet Transform (ShIWT) and the Nonlinear Energy Operator (NEO). In our method, ShIWT was used to reduces the noise sources that affect the ECG signal. At the second stage, NEO was applied to highlight the QRS complex patterns. Finally, a simple thresholding function and efficient decision rules were applied for detecting R peak positions. The performance of the proposed method is evaluated over MIT-BIH Arrythmia Database (MITDB) [6]. In comparison with the state of the art methods, the proposed algorithm achieves a satisfactory and comparable performances.

The rest of this paper is organized as follows. Section 2 presents the ECG dataset as well as the methods (i.e. Shift Invariant Wavelet Transform and Nonlinear Energy Operator) used in this work. In Sect. 3, we explain the main steps of our proposed approach. Performances evaluation of the proposed method was achieved in Sect. 4. Finally, Sect. 5 concludes this work.

2 Materials and Methods

2.1 ECG Dataset: MIT-BIH Arrhythmia Database (MITDB)

In order to validate the efficiency of the developed method, real ECG signals taken from the standard MITDB were used. This database contains 48 ECG recordings with a 30 min time duration sampled at 360 Hz with a resolution of 11 bits. The majority of ECG records are recorded with the Modified Limb Lead II (ML-II). Each ECG signal from this database is annotated by two experts cardiologists which have determined the QRS complex time positions.

2.2 Shift Invariant Wavelet Transform

Wavelet Transform is a useful signal processing tool that has been applied to different research fields. In this study, we use ShIWT [7] dues to its shift invariance property. ShIWT decomposes an ECG signal $x[n]$ into different frequency sub-bands by passing $x[n]$ through a Wavelet filter bank composed by high pass $G_j[n]$ and low pass $H_j[n]$ filters. Output coefficients of $H_j[n]$ are known as approximations coefficients (A_j) while those of $G_j[n]$ are known as details ones (D_j). ShIWT up-samples the coefficients of the low pass and high pass filters in such a way that its outputs at each level have the same length as the original signal [8]. ShIWT process at level $j + 1$ can be expressed as:

$$A_{j+1}[k] = \sum_n A_j[n]H_j[k - n] \tag{1}$$

$$D_{j+1}[k] = \sum_n A_j[n]G_j[k - n] \tag{2}$$

where $A_0[n]$ is the original signal $x[n]$.

2.3 Nonlinear Energy Operator (NEO)

Teager Energy Operator (TEO) is originally developed by Kaiser [9] in order to measure the physical energy required to generates a signal. In its discrete form, TEO for a signal $x[n]$ can be computed as the following equation:

$$\psi_x[n] = x^2[n] - x[n - 1] * x[n + 1] \tag{3}$$

TEO has been widely used in many research fields such as spike detection in EEG signal. However, TEO is sensitive to the noise components. To overcome this drawback, Nonlinear Energy Operator (NEO) defined as smoothed version of TEO has been proposed by Mukhopadhyay and Ray [10] by convolving $\psi_x[n]$ with a time domain window $w[n]$ as follows:

$$\Psi_s[n] = \psi_x[n] \otimes w[n] \tag{4}$$

where $\otimes$ denotes the convolution operator and $w[n]$ represents the time window. By analyzing Eqs. (3) and (4), it is clearly viewed that NEO can be applied to detect the spiky activities such as QRS complex in an ECG signal.

3 Proposed QRS Detector

A schematic diagram of the principal steps of our proposed method is illustrated in Fig. 1. It includes three key stages that will be detailed in the following subsections.

3.1 ECG Pre-processing Based Wavelet

In this study, ECG filtering based on ShIWT is performed as follows:

1. Apply ShIWT to the original ECG signal $ecg[n]$ using '$coif1$' wavelet.

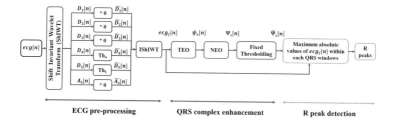

Fig. 1. Flowchart of the proposed method

2. Thresholding detail coefficients at levels 4 and 5 which contain the useful information of QRS wave with a global frequency band of [5.625–22.5 Hz].
3. Zeroing detail coefficients at levels 1, 2, 3 and the approximation coefficient at level 5 to filter out higher and lower frequency noises respectively.
4. Apply Inverse ShIWT to obtain the filtered ECG signal $ecg_f[n]$.

In our scheme, we have used the soft thresholding function defined as [11]:

$$\tilde{D}_j[n] = \begin{cases} sgn(D_j[n])(\mid D_j[n] \mid -Th_j) & ; \quad \mid D_j[n] \mid \geq Th_j \\ 0 & ; \quad \mid D_j[n] \mid < Th_j \end{cases} \tag{5}$$

where $sgn(.)$ is the sign function, Th_j is the threshold at the jth level and $\tilde{D}_j[n]$ is the thresholded version of noisy detail $D_j[n]$. Th_j is derived from [12] as follows:

$$Th_j = K\sigma_j\sqrt{2logN}, \quad j = 4, 5 \tag{6}$$

where K is a constant value ($K = 0.6$ in this work), N is the signal length and $\sigma_j = MAD(D_j)/0.6745$ is the estimated noise variance at the jth detail.

3.2 QRS Complex Enhancement Using NEO

Subsequently to obtain the filtered ECG signal, TEO and NEO were applied respectively to $ecg_f[n]$ in order to enhance QRS complex patterns of an ECG signal, and to reduce the effect of unwanted peaks such P and T waves of this signal. To illustrate this observation, an example of application is shown in Fig. 2 when the traditional squaring operation $E[n] = ecg^2[n]$, TEO and NEO (using Bartlett Window with 160 ms width, which corresponds to the largest QRS complex [13]) are applied to time segment from '113.dat' ECG record.

According to Fig. 2, we notice that NEO is able to enhances the QRS complex, and to reduces the slowly parts in contrast to the squaring operation when T waves are also highlighted as shown in Fig. 2(b). After amplifying QRS features, NEO output will be thresholded to retain only the candidate QRS complex emplacements. Hence, as used in the literature of spike detection based on NEO, the threshold τ was computed as a scaled version of the mean over the entire NEO output as follows:

$$\tau = \beta\frac{1}{N}\sum_{n=1}^{N}\Psi_s[n] \tag{7}$$

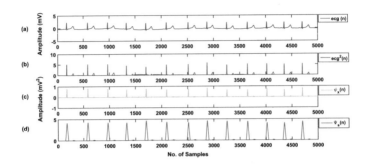

Fig. 2. Results of Squaring operation, TEO and NEO operators for an ECG signal: (a) ECG signal $x[n]$, (b) Squaring operation output $E[n]$, (c) TEO output $\psi_x[n]$ and (d) NEO output $\Psi_s[n]$.

where N is the signal length and β is a tuning parameter that is determined empirically. In this work, β was fixed to 35%. Hence, all samples of NEO output below τ will be set to zero to obtain a thresholded version $\tilde{\Psi}_s[n]$.

3.3 R Peak Detection Algorithm

Finally, R peak time positions can be detected using a sliding window based strategy over the entire $\tilde{\Psi}_s[n]$ scanned with the filtered ECG signal $ccg_f[n]$. Hence, the maximum absolute value of $ecg_f[n]$ within each QRS window is considered to be the truth R peak. Furthermore, to reduce the detected false R peaks, we apply two decision rules: (1) The standard rule based on the refractory period of 200 ms (i.e. 72 samples) within a new QRS complex cannot occur was applied to detect each new R peak; (2) If two consecutive R peaks were detected in a time interval smaller than 250 ms (i.e. 90 samples), we retain only the peak which having the maximal absolute value in $ecg_f[n]$.

4   Tests and Results

4.1 Performance Evaluation Criteria

In order to evaluate our approach, the Sensitivity (Se), the Positive Predictivity (P_+) and the Detection Error Rate (DER) are calculated respectively as the following equations:

$$Se = \frac{TP}{TP + FN} \tag{8}$$

$$P_+ = \frac{TP}{TP + FP} \tag{9}$$

$$DER = \frac{FN + FP}{TB} \tag{10}$$

where TP (True Positive or correct peaks), FP (False Positive or false peaks), FN (False Negative or missing peaks) and TB is the total number of beats.

4.2   Experimental Results

The overall performances achieved by the proposed method, in terms of the three evaluation criteria, are reported in Table 1. As illustrated in this table, the proposed QRS detector was successful to detect correctly 109230 beats from a total number of 109494 beats with a sensitivity of $Se = 99.76\%$, positive predictivity $P_+ = 99.77\%$ and detection error rate of $DER = 0.47\%$.

Figure 3 shows the different stages of our proposed approach applied to time segment taken from ECG record '108.dat'. In this figure, Fig. 3(a) shows the original ECG signal $ecg[n]$, Fig. 3(b) illustrates the filtered signal $ecg_f[n]$, Fig. 3(c) depicts the thresholded version $\tilde{\Psi}_s[n]$ of NEO output $\Psi_s[n]$. The detected R peaks are shown in Fig. 3(f) when the peak time positions are scanned with the annotated QRS complex from MITDB. This figure shows that all R peaks are truly detected by the proposed method.

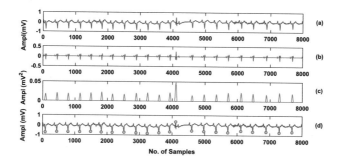

Fig. 3. Results of QRS detection (a) ECG data $ecg[n]$, (b) Filtered signal $ecg_f[n]$, (c) Thresholded version $\tilde{\Psi}_s[n]$ of NEO output $\Psi_s[n]$, (d) ECG signal overlaid by the markers from the proposed algorithm (red circle) and with MITDB annotations (green circle).

4.3   Comparative Analysis with the Literature

Table 2 presents the results obtained by our approach against some recently developed techniques. By analyzing the reported results in this table, it is clearly observed that our approach achieves satisfactory performances compared to some established algorithms. Unfortunately, as we can see in Table 1, the proposed QRS detector provides higher DER values for some ECG recordings such as those numbered '104.dat', '105.dat', '203.dat', '207.dat', '208.dat' and '228.dat'. In these records, ECG data are affected by various types of noises and different QRS morphologies which make R peak detection very difficult.

Table 1. Experimental results of the proposed method on MITDB

ECG	TB	NDB	TP	FP	FN	Se(%)	P_+(%)	DER(%)
100	2273	2273	2273	0	0	100.00	100.00	0.000
101	1865	1869	1864	5	1	99.95	99.73	0.322
102	2187	2187	2187	0	0	100.00	100.00	0.000
103	2084	2084	2084	0	0	100.00	100.00	0.000
104	2229	2262	2224	38	5	99.78	98.32	1.929
105	2572	2605	2556	49	16	99.38	98.12	2.527
106	2027	2011	2010	1	17	99.16	99.95	0.888
107	2137	2135	2134	1	3	99.86	99.95	0.187
108	1763	1762	1749	13	14	99.21	99.26	1.531
109	2532	2523	2523	0	9	99.64	100.00	0.355
111	2124	2123	2123	0	1	99.95	100.00	0.047
112	2539	2540	2539	1	0	100.00	99.96	0.039
113	1795	1824	1795	29	0	100.00	98.41	1.616
114	1879	1887	1879	8	0	100.00	99.58	0.426
115	1953	1953	1953	0	0	100.00	100.00	0.000
116	2412	2391	2388	3	24	99.00	99.87	1.119
117	1535	1535	1535	0	0	100.00	100.00	0.000
118	2278	2279	2278	1	0	100.00	99.96	0.044
119	1987	1988	1987	1	0	100.00	99.95	0.050
121	1863	1862	1862	0	1	99.95	100.00	0.054
122	2476	2476	2476	0	0	100.00	100.00	0.000
123	1518	1519	1518	1	0	100.00	99.93	0.066
124	1619	1619	1619	0	0	100.00	100.00	0.000
200	2601	2605	2599	6	2	99.92	99.77	0.308
201	1963	1957	1952	5	11	99.44	99.74	0.815
202	2136	2134	2134	0	2	99.91	100.00	0.094
203	2980	2957	2938	19	42	98.59	99.36	2.047
205	2656	2653	2653	0	3	99.89	100.00	0.113
207	1860	1866	1852	14	8	99.57	99.25	1.183
208	2955	2929	2923	6	32	98.92	99.80	1.286
209	3005	3005	3005	0	0	100.00	100.00	0.000
210	2650	2642	2636	6	14	98.47	99.77	0.755
212	2748	2749	2748	1	0	100.00	99.96	0.036
213	3251	3244	3241	3	10	99.69	99.91	0.400
214	2262	2258	2256	2	6	99.73	99.91	0.354
215	3363	3362	3360	2	3	99.91	99.94	0.149
217	2208	2204	2203	1	5	99.77	99.95	0.272
219	2154	2154	2154	0	0	100.00	100.00	0.000
220	2048	2048	2048	0	0	100.00	100.00	0.000
221	2427	2424	2424	0	3	99.88	100.00	0.124
222	2483	2482	2472	10	11	99.56	99.60	0.846
223	2605	2602	2602	0	3	99.88	100.00	0.115
228	2053	2068	2045	23	8	99.61	98.89	1.510
230	2256	2258	2256	2	0	100.00	99.91	0.089
231	1571	1571	1571	0	0	100.00	100.00	0.000
232	1780	1783	1780	3	0	100.00	99.83	0.169
233	3079	3071	3071	0	8	99.74	100.00	0.260
234	2753	2752	2752	0	1	99.96	100.00	0.036
All	**109494**	**109484**	**109230**	**254**	**264**	**99.76**	**99.77**	**0.473**

Table 2. Quantitative comparison of the proposed method with the literature.

QRS detection algorithm	Se(%)	P_+(%)	DER(%)
Merah et al. [14]	99.84	99.88	0.28
Bouaziz et al. [5]	99.87	99.79	0.34
ShIWT-NEO (this work)	*99.76*	*99.77*	*0.47*
Banerjee et al. [13]	99.60	99.50	0.61
Yochum et al. [15]	99.85	99.48	0.67

5 Conclusion

In this paper, we have presented an efficient simple tool for QRS complex detection in ECG signal. The proposed detector is based on Shift Invariant Wavelet Transform and Nonlinear Energy Operator. The proposed approach is validated on the whole known MITDB. The overall results show that our method is able to gives a comparable or higher performances against existing methods. However, the important parameters of the proposed algorithm should be optimized which can improve certainly the accuracy of R peak detection.

References

1. Kohler, B., Hennig, C., Orglmeiste, R.: The principles of software QRS detection. In: IEEE Engineering in Medicine and Biology Society (EMBC), pp. 42–57 (2002)
2. Pan, J., Tompkins, W.: A real time QRS detection algorithm. IEEE Trans. Biomed. Eng. **3**(32), 230–236 (1985)
3. Hu, Y.H., Tompkins, W., Urrusti, J., Afonso, V.: Applications of artificial neural networks for ECG signal detection and classification. J. Electrocardiol. **26**, 66–73 (1993)
4. Benitez, S., Gaydecki, P., Zaidi, A., Fitzpatrick, A.: The use of the Hilbert transform in ECG signal analysis. Comput. Biol. Med. **31**, 399–406 (2001)
5. Bouaziz, F., Boutana, D., Benidir, M.: Multiresolution wavelet-based QRS complex detection algorithm suited to several abnormal morphologies. IET Signal Process. **8**(7), 774–782 (2014)
6. Mark, R., Moody, G.: MIT-BIH-Arrhythmia Database. http://www.physionet.org/physiobank/database/mitdb
7. Nason, G., Silverman, B.: The stationary wavelet transform and some statistical applications. In: Antoniadis, A., Oppenheim, G. (eds.) Wavelets and Statistics. Lecture Notes in Statistics, pp. 281–299 (1995)
8. El Bouny, L., Khalil, M., Adib, A.: Performance analysis of ECG signal denoising methods in transform domain. In: The Third IEEE ISCV, Fes, Morocco, April 2018
9. Kaiser, J.: On a simple algorithm to calculate the energy' of a signal. In: IEEE ICASSP, pp. 381–384 (1990)
10. Mukhopadhyay, S., Ray, G.C.: A new interpretation of nonlinear energy operator and its efficacy in spike detection. IEEE Trans. Biomed. Eng. **45**(2), 180–187 (1998)

11. Donoho, D.: De-noising by soft-thresholding. IEEE Trans. Inf. Theory **41**(3), 613–627 (1995)
12. Johnstone, I., Silverman, B.: Wavelet threshold estimators for data with correlated noise. J. R. Stat. Soc. Ser. B (Gen.) **59**, 319–351 (1997)
13. Banerjee, S., Gupta, R., Mitra, M.: Delineation of ECG characteristic features using multiresolution wavelet analysis method. Measurement **45**, 474–487 (2012)
14. Merah, M., Abdelmalik, T., Larbi, B.: R-peaks detection based on stationary wavelet transform. Comput. Methods Prog. Biomed. **121**, 1–12 (2015)
15. Yochum, M., Renaud, C., Jacquir, S.: Automatic detection of P, QRS and T patterns in 12 leads ECG signal based on CWT. Biomed. Signal Process. Control **25**, 46–52 (2016)

New Faster R-CNN Neuronal Approach for Face Retrieval

Imane Hachchane[1]([✉]), Abdelmajid Badri[1], Aïcha Sahel[1],
and Yassine Ruichek[2]

[1] Laboratoire d'Electronique, Energie, Automatique & Traitement de
l'Information (EEA&TI), Faculté des Sciences et Techniques Mohammedia,
Université Hassan II Casablanca, B.P. 146, 20650 Mohammedia, Morocco
`hachchaneimane@gmail.com,`
`abdelmajid_badri@yahoo.fr, sahel_ai@yahoo.fr`
[2] IRTES-Laboratoire SET, Université de Technologie de Belfort Montbéliard,
Belfort, France
`yassine.ruichek@utbm.fr`

Abstract. Image features acquired from pre-entrained convolutional neural networks (CNN) are becoming the norm in instance retrieval. This work investigates the relevance of using an object detection CNN like Faster R-CNN as a feature extractor. We built a face retrieval pipeline composed of filtering and re-ranking that uses the objects proposals learned by a Region Proposal Network (RPN) and their associated representations taken from a CNN. Moreover, we study the relevance of Faster R-CNN representations when the network is fine-tuned for the same data that we want to recover. We evaluate the performance of the system with the Labeled Faces in the Wild (LFW) database 13k, the FERET database 3K, and Faces94 database 2k. The results obtained are very encouraging.

Keywords: Image processing · Classification · Object recognition · CNN ·
Faster R-CNN · Image instance search · Face retrieval

1 Introduction

Visual search applications have gained substantial popularity recently due to the explosion of visual content that we are witnessing nowadays. This proliferation of visual media motivated researchers to offer effective content-based image retrieval systems. This work addresses the problem of instance search but more specifically face retrieval, understood as the task of recovering query faces from a database containing instances of those queries.

Convolutional neural networks (CNNs) achieved state-of-the-art results in many visual tasks such as pedestrian detection, image classification [20] and specially face recognition [30]. Chandrasekhar proposed using Convolutional Neural Nets (CNN) representations as a global descriptor for image retrieval [28]. CNNs trained with large amounts of data can learn features generic enough to be used to solve tasks for which the network has not been trained [16]. For image retrieval, in particular, many works in the literature [3, 23] have adopted solutions based on standard features

© Springer Nature Switzerland AG 2019
F. Khoukhi et al. (Eds.): AIT2S 2018, LNNS 66, pp. 113–120, 2019.
https://doi.org/10.1007/978-3-030-11914-0_12

extracted from a pre-established CNN for image classification [11, 20, 22], achieving encouraging performances.

Content-based image retrieval systems usually combine a filtering step, where all the images in a database are sorted according to their similarity to the query image, with more expensive computationally mechanisms that are only applied to the most relevant items. Geometric verification and spatial analysis [9, 13, 18, 21] are common redistribution strategies, which are often followed by a query expansion (pseudo-relevance feedback) [1, 5].

Spatial re-ranking usually indicates the use of sliding windows of different scales and aspect ratios on an image. Each sliding window is compared to the query instance to find the optimal location that contains the query, requiring the calculation of a visual descriptor on each of the windows considered [26]. Such a strategy resembles that of an object detection algorithm, which generally evaluates many image locations and determines whether or not the object is there. Nowadays the usage of exhaustive search with sliding windows or the computation of object proposals [2, 24] is no longer required, CNN's [17] are used end-to-end to simultaneously learn the locations of objects and their labels.

This work explores the relevance of the standard and fine-tuned features of an object detection CNN for the face retrieval task.

2   Related Work

Convolutional Neural Networks for Instance Search. Early works using features from pre-trained image classification CNN's, showed that using fully connected layers for image retrieval were more suitable [4]. Razavian et al. [16], combined fully connected layers extracted from different image sub-much witch improved the results. Later on, a new generation of works explored using of other layers of pre-trained CNN and found that convolutional layers significantly outperform fully connected layers during image recovery tasks [19, 26].

Convolutional Neural Networks for Object Detection. Many CNN-based object detection pipelines have been proposed. Girshick et al. Presented R-CNN [8], a version of AlexNet by Krizhevsky [11], refined for Pascal VOC Detection [6]. Girshick later published Fast R-CNN [7], adopting the same speed strategy as SPP-net, but largely, it replaces the post-hoc training of SVM classifiers and box regressors with a solution that allows training to be carried out end-to-end. Since then Ren et al. Introduced the Faster R-CNN [17], which introduced a Region Proposal Network (RPN) that removes the dependence of object proposals from older CNN object detection systems. In Faster R-CNN, RPN shares features with the object-detection network in [7] to simultaneously learn prominent object propositions and their associated class probabilities. Although the Faster R-CNN is designed for generic object detection, Jiang et al. [29] demonstrated that it can achieve impressive face detection performance specially when retrained on a suitable face detection training set.

In this work we use the convoluted features of a state of the art pre-trained object detection CNN Faster R-CNN. We exploit his end-to-end autonomous object detection architecture to extract image-based and region-based convolutional features in a single forward pass and test their relevance for face retrieval.

3 Methodology

3.1 CNN-Based Representations

We explore the relevance of using features extracted from an object detection CNN for the face retrieval task. We are using query instances defined by a bounding box above the query image. We select Faster R-CNN architecture and its pre-trained models [17] and use them as a global and local feature extractor. Faster R-CNN is composed of a region proposal network that learns a set of windows locations, and a classifier that learns to labels each of those windows as one of the classes in the learning dataset [26].

As with other works [3, 23, 26], our goal is to extract a representation of compact images constructed from activations of a convolutional layer in a CNN. Faster R-CNN works faster on a global and local scale. We build a global image descriptor by ignoring all the layers of Faster R-CNN that works with object proposals and extract features from the last convolutional layer. Considering the extracted activations of a convolution layer for an image, we group the activations of each filter to create an image descriptor with the same dimension as the number of filters in the convolution layer. We aggregate the activations of each window suggestion in the RoI Pooling layer to create regional descriptions.

3.2 Fine-Tuning Faster R-CNN

The relevance of this technique based on the Faster R-CNN method allows us to (1) obtain better representations of features for face retrieval and (2) to improve the performance of spatial analysis and re-ranking. To achieve this, we choose to fine-tune Faster RNN to detect the query faces to be retrieved by our system. For this purpose, we modify the architecture of Faster R-CNN to display the coordinates of the bounding box and the class scores for each of the query instances of the tested instances.

The resulting refined networks must be used to extract better representations of images and regions and to perform spatial redirection based on class scores instead of entity similarities.

3.3 Image Recovery

This section describes the three steps of the face retrieval pipeline:

Filtering Step. We will create image descriptors for query and database images. At testing time, the descriptor of the query image is compared to all items in the database, which are then ranked according to the similarity of the cosine. At this stage, the entire image is considered as a query.

Spatial Re-ranking. After the filtering step, the N upper elements are analyzed locally and re-ranked.

Query Expansion (QE). We average of the image descriptors of the M higher elements of the ranking with query descriptor to carry out a new search.

4 Experiments

4.1 Datasets Exploited

We evaluate our methodologies using the following datasets:

- LFW [14]: 13233 images, including 55 Query images. A framing box surrounding the target face is provided for query images.
- FERET [15]: 3528 images, including 55 Query images. A framing box surrounding the target face is provided for query images.
- FACES94 [27]: 2809 images, including 55 Query images. A framing box surrounding the target face is provided for query images.
- Facescrub [10]: 55127 images.

4.2 Experimental Setup

We use both the VGG16 [20] and the ZF [25] architecture of Faster R-CNN to extract the global and local features. In both cases, we use the last convolution layer (conv5 and conv5_3 for ZF and VGG16, respectively) to create the global descriptors which are of dimension 256 and 512 for the ZF and VGG16 architectures, respectively.

The local features are grouped from the Faster R-CNN RoI clustering layer. The images are resized so that their smallest size is 600 pixels. All experiments were performed on a Nvidia GTX GPU.

4.3 Off-the-Shelf Faster R-CNN Features

In this section, we evaluate the performance of using features extracted from Faster R-CNN for face retrieval.

We carried out a comparative study of the sum and max-pooling strategies of the image-wise and region-wise descriptors. Table 1 summarizes the results. According to our experiments, the max-pooling is significantly higher than the sum pooling for the filtering phase. It also shows the performance of different Faster R-CNN architectures (ZF and VGG16) trained on two different datasets (Pascal VOC and COCO [12]), including query expansion experiments with M = 5 images retrieved. As expected, the features extracted from the VGG16 network performed better in most cases, which is consistent with previous work in the literature showing that the capabilities of deeper networks achieve better performance. Query expansion applied after spatial re-ranking results in significant gains for most tested datasets.

Table 1. Mean Average Precision (MAP) of pre-trained Faster R-CNN models with ZF and VGG16 architectures. (P) and (C) denote whether the network was trained with Pascal VOC or Microsoft COCO images, respectively. With a comparison between sum and max pooling strategies. When indicated, QE is applied with M = 5.

NET	QE	Pooling	LFW	Faces94	FERET
ZF (P)	Yes	Sum	0.1573	0.7295	**0.5171**
		Max	0.1321	**0.9517**	0.4018
	No	Sum	0.1647	0.6334	0.5014
		Max	0.1370	0.9494	0.4061
VGG16 (P)	Yes	Sum	0.1493	0.7328	0.4666
		Max	0.1595	0.9466	0.4269
	No	Sum	0.1629	0.6317	0.4386
		Max	**0.1648**	0.9431	0.4240
VGG (C)	Yes	Sum	0.1528	0.5895	0.3126
		Max	0.1409	0.9518	0.4126
	No	Sum	0.1576	0.4800	0.3126
		Max	0.1628	0.9497	0.4129

4.4 Fine-Tuning Faster R-CNN

In this part, we evaluate the impact of fine-tuning a pre-trained network on recovery performance with the query objects to retrieve. We chose to refine the model VGG16 Faster R-CNN, pre-trained with the objects of Pascal VOC.

For the FERET and Faces94 datasets, we modify the output layer in the network to return 422 class probabilities (269 people in the FERET dataset plus 152 people in the Faces94 dataset, plus one additional class for the background) and their corresponding bounded bound box coordinates. We use the 6337 images of both databases and we create their delimiting locations and use them as training data for our first fine-tuned network which we called VGG (FF).

For the LFW we train the network on a similar database called Facescrub. We modify the output layer in the network to return 530 class probabilities (530 people, plus one additional class for the background) and their corresponding bounded bound box coordinates. We use the 55127 images of the database and their delimiting locations as training data for our second fine-tunes network which we called VGG (LFW).

The Faster R-CNN original parameters described in [17] are kept, except for the number of iterations, which we decreased to 20,000 instead of 80,000 given our smaller number of training samples.

We then use the refined networks of the tuning strategy (VGG (FF) & VGG (LFW)) on all datasets to extract image and region descriptors to perform a face retrieval. Table 2 presents the results obtained. As expected, the refined features significantly exceed the raw Faster R-CNN features for all data sets (mAP is ∼75%, ∼1.2%, and ∼72% higher for LFW, Faces94, and FERET, respectively) (Fig. 1 and Table 3).

Table 2. Results obtained when using the fine-tuned models. When indicated, QE is applied with m = 5.

NET	QE	LFW	Faces94	FERET
VGG16 (FF)	Yes	0.1593	**0.9638**	**0.8913**
	No	0.1665	**0.9638**	0.8698
VGG16 (LFW)	Yes	**0.2893**	0.9492	0.6027
	No	0.2864	0.9430	0.5936

Table 3. Training and testing time.

	FERET	Faces94	LFW
Training faster R-CNN models	2 h 47 min		2 h30 min
Feature extraction	8 min	5 min	21 min
Ranking/Reranking	30 min	34 min	15 min

Fig. 1. Result example of the fine-tuned network

5 Conclusion

This article presents a strategy for using the CNN features of an object detection CNN. It provides a simple baseline that uses the Faster R-CNN features to describe the images and their subparts. We have shown that it is possible to improve the performance of a system by using image and region features from the Faster R-CNN Object Detection CNN. We take advantage of the proposals of objects learned by a PNR (Region Proposal Network) and their associated features taken from a CNN to build a face retrieval pipeline. Moreover, we have shown the importance of Faster R-CNN features when the network is fine-tuned for the same objects that we want to recover. The results obtained are encouraging. In future work, we intend to make a comparison with YOLO algorithm to see which one perform better in face retrieval.

References

1. Arandjelovic, R., Zisserman, A.: Three things everyone should know to improve object retrieval. In Computer Vision and Pattern Recognition (CVPR), pp. 2911–2918 (2012)
2. Arbeláez, P., Pont-Tuset, J., Barron, J., Marques, F., Malik, J.: Multiscale combinatorial grouping. In: Proceedings of the IEEE Conference on Computer Vision and Pattern Recognition, pp. 328–335 (2014)
3. Babenko, A., Lempitsky, V.: Aggregating local deep features for image retrieval. In: International Conference on Computer Vision (ICCV), December 2015
4. Babenko, A., Slesarev, A., Chigorin, A., Lempitsky, V.: Neural codes for image retrieval. In: Computer Vision–ECCV 2014, pp. 584–599 (2014)
5. Chum, O., Philbin, J., Sivic, J., Isard, M., Zisserman, A.: Total recall: automatic query expansion with a generative feature model for object retrieval. In: International Conference on Computer Vision, pp. 1– 8 (2007)
6. Everingham, M., Van Gool, L., Williams, C.K.I., Winn, J., Zisserman, A.: The pascal visual object classes (voc) challenge. Int. J. Comput. Vis. **88**(2), 303–338 (2010)
7. Girshick, R.: Fast R-CNN. In: Proceedings of the IEEE International Conference on Computer Vision (2015)
8. Girshick, R., Donahue, J., Darrell, T., Malik, J.: Rich feature hierarchies for accurate object detection and semantic segmentation. In: Proceedings of the IEEE Conference on Computer Vision and Pattern Recognition, pp. 580–587 (2014)
9. Jégou, H., Douze, M., Schmid, C.: Improving bag-of-features for large scale image search. Int. J. Comput. Vis. **87**(3), 316–336 (2010)
10. Ng, H.-W., Winkler, S.: A data-driven approach to cleaning large face datasets. In: Proceedings of IEEE International Conference on Image Processing (ICIP), Paris, France (2014)
11. Krizhevsky, A., Sutskever, I., Hinton, G.E.: Imagenet classification with deep convolutional neural networks. In: Advances in Neural Information Processing Systems, pp. 1097–1105 (2012)
12. Lin, T.-Y., Maire, M., Belongie, S., Hays, J., Perona, P., Ramanan, D., Dollár, P., Zitnick, C. L.: Microsoft coco: common objects in context. In: Computer Vision–ECCV 2014, pp. 740–755. Springer (2014)
13. Mei, T., Rui, Y., Li, S., Tian, Q.: Multimedia search re-ranking: a literature survey. ACM Comput. Surv. (CSUR) **46**(3), 38 (2014)
14. Huang, G.B., Ramesh, M., Berg, T., Learned-Miller, E.: Labeled faces in the wild: a database for studying face recognition in unconstrained environments. University of Massachusetts, Amherst
15. Jonathon Phillips, P., Wechsler, H., Huang, J., Rauss, P.J.: The FERET database and evaluation procedure for face-recognition algorithms. Image Vis. Comput. **16**(5), 295–306 (1998)
16. Razavian, A.S., Azizpour, H., Sullivan, J., Carlsson, S.: CNN features off-the-shelf: an astounding baseline for recognition. In: Computer Vision and Pattern Recognition Workshops (CVPRW) (2014)
17. Ren, S., He, K., Girshick, R., Sun, J.: Faster R-CNN: towards real-time object detection with region proposal networks. In: Advances in Neural Information Processing Systems, pp. 91–99 (2015)
18. Zhang, W., Ngo, C.-W.: Topological spatial verification for instance search. IEEE Trans. Multimed. **17**, 1236–1247 (2015)

19. Sharif Razavian, A., Sullivan, J., Maki, A., Carlsson, S.: A baseline for visual instance retrieval with deep convolutional networks. In: International Conference on Learning Representations, ICLR (2015)
20. Simonyan, K., Zisserman, A.: Very deep convolutional networks for large-scale image recognition. CoRR, abs/1409.1556 (2014)
21. Zhang, Y., Jia, Z., Chen, T.: Image retrieval with geometry- preserving visual phrases. In: Computer Vision and Pattern Recognition (CVPR), pp. 809–816 (2011)
22. Szegedy, C., Liu, W., Jia, Y., Sermanet, P., Reed, S., Anguelov, D., Erhan, D., Vanhoucke, V., Rabinovich, A.: Going deeper with convolutions. In: Proceedings of the IEEE Conference on Computer Vision and Pattern Recognition, pp. 1–9 (2015)
23. Tolias, G., Sicre, R., Jégou, H.: Particular object retrieval with integral max-pooling of CNN activations. In: ICLR (2016)
24. Uijlings, J.R., van de Sande, K.E., Gevers, T., Smeulders, A.W.: Selective search for object recognition. Int. J. Comput. Vis. 104(2), 154–171 (2013)
25. Zeiler, M.D., Fergus, R.: Visualizing and understanding convolutional networks. In: Computer vision–ECCV 2014, pp. 818–833. Springer (2014)
26. Salvador, A., Giro-I-Nieto, X., Marques, F., Satoh, S.I.: Faster R-CNN features for instance search. In: Proceedings of the IEEE Conference on Computer Vision and Pattern Recognition Workshops (2016)
27. Spacek, L.: Faces94 a face recognition dataset (2007)
28. Chandrasekhar, V., Lin, J., Morère, O., Goh, H., Veillard, A.: A practical guide to CNNs and fisher vectors for image instance retrieval. Signal Process. **128**, 426–439 (2016)
29. Jiang, H., Learned-Miller, E.: Face detection with the faster R-CNN. In: Proceedings - 12th IEEE International Conference on Automatic Face Gesture Recognition, FG 2017 - 1st International Workshop Adaptive Shot Learning for Gesture Understanding Production, ASL4GUP 2017, Biometrics Wild, Bwild 2017, Heterogeneous, pp. 650–657 (2017)
30. Sun, Y., Wang, X., Tang, X.: Deep learning face representation from predicting 10,000 classes. In: Computer Vision and Pattern Recognition (CVPR), pp. 1891–1898 (2014)

An Efficient Level Set Speed Function Based on Temperature Changes for Brain Tumor Segmentation

Abdelmajid Bousselham$^{(\boxtimes)}$, Omar Bouattane, Mohamed Youssfi, and Abdelhadi Raihani

Laboratory SSDIA, ENSET Mohammedia, University Hassan 2 Casablanca, Casablanca, Morocco
abdelmajid.bousselham@gmail.com,
o.bouattane@gmail.com, med@youssfi.net,
abraihani@yahoo.fr

Abstract. In clinical routine, accurate segmentation of brain tumors from Magnetic Resonance Images (MRI) plays an important role in diagnostic; it is a challenging and difficult task as brain tumors have various appearance properties. In this study, a modified level set speed function for accurate brain tumor segmentation applied on thermal images to reinforce brain tumors segmentation using MRI is presented. Tumor cells have high temperature compared to healthy cells, due to the high metabolic activity of abnormal cells. To calculate the thermal image we have used Pennes BioHeat Transfer Equation (PBHTE) resolved using Finite Difference Method (FDM). By analyzing the tumor thermal profile, the temperature is higher in the tumor center and is reduced as we move to the tumor borders; we have used this physical phenomenon in level set function for tumor segmentation. The proposed approach is tested in synthetic MRI images containing tumors with different volumes and locations. The obtained results showed that 10.29 % of brain tumor segmented correctly by level set method in the thermal image as a tumor part, contrarily in T1 which is segmented as healthy tissue, the same for T1c and Flair with 4.32 % and 22.58 % respectively. Therefore, the temperature can play an important role to improve the accuracy of brain tumor segmentation in MRI.

Keywords: Magnetic Resonance Imaging · Bioheat Transfer · Finite Difference Method · Level set method

1 Introduction

Brain tumor segmentation using MRI images represents a challenging task in diagnosis. Several MRI sequences can be obtained; they are called weighted images, such as T1-weighted, T2-weighted, Proton-Density Weighted, Fluid-attenuated Inversion Recovery (FLAIR), etc. T1-weighted is suitable for segmentation of brain tissues as it shows a good contrast between gray matter and white matter [1], T1-weighted contrast-enhanced, T2-weighted, and Flair are the most used MRI sequences for brain tumors structures segmentation as it makes tumor region hyperintense. In this paper, to validate

© Springer Nature Switzerland AG 2019
F. Khoukhi et al. (Eds.): AIT2S 2018, LNNS 66, pp. 121–129, 2019.
https://doi.org/10.1007/978-3-030-11914-0_13

our approach for brain tumor segmentation, we have used synthetic T1, T1 contrast, T2, and Flair MRI images.

In medical imaging, segmentation is an active field in clinical routine, which consists in extracting from the image one or more regions forming the area of interest. Various approaches have been proposed in the literature to perform brain tumor detection and extraction, among them we find, threshold-based methods, region-based methods, deformable methods, classification methods, and deep learning [2–6]. Deformable models are popular methods for brain tumor segmentation in MRI. They are modeled by curves (2D) or surfaces (3D) initialized in an image and move by the influence of internal and external forces. Internal or local forces are used to keep the curve smooth during the deformation process, external forces are calculated from image pixels to move the model towards the boundaries of the object of interest. There is to types of deformable models, parametric models or snakes [7] and geometric models or level sets. The parametric models need a parametric representation during propagation of the curve, the limitation of these models is that the curve cannot split and merge in order to segment multiple objects. The second type of deformable models are level sets [8], they move the model using geometric measurements like the normal and curvature of the curve. The main advantage of level sets method is it can implicitly segment multiple objects.

Accurate segmentation of brain tumors represents a challenging task since the tumors have various appearance properties [4]. The main contribution of the present work is to improve the segmentation algorithms accuracy using thermal analysis of tumors. Brain tumor represents a heat source and can be used to detect the presence of a tumor and determine its size and location (thermography) [9–11]. In this work, we have applied level set method in the thermal image to determine tumor contours, we used Pennes bioheat equation for calculating thermal images, as in our knowledge there is no MRI thermal image sequence; we compared the obtained results with the level set method applied in different MRI sequences.

The rest of the study presents the proposed approach in Sect. 2, obtained results and discussion in Sect. 3. Finally, the conclusion in Sect. 4.

2   Methods

2.1   Temperature Calculation

Thermal images are calculated using Pennes bioheat equation was used [12, 13], it is a partial differential equation described as follows:

$$\rho C_P \frac{\partial T}{\partial t} = K \cdot \left(\frac{\partial^2 T}{\partial x^2} + \frac{\partial^2 T}{\partial y^2} \right) + \omega_b \rho_b C_{pb}(T_a - T) + Q_m, \tag{1}$$

where $\rho [\text{Kg/m}^3]$ is the density of tissue, C_P [J/(Kg °C)] is the specific heat of tissue, is the thermal conductivity, ω_b[ml /(s · ml)] is the blood perfusion rate, ρ_b [Kg/m³] is the density of blood, C_{pb} [J/(Kg °C)] is the specific heat of blood, T_a [°C] is the temperature of artery and Q_m [W/m³] is metabolic heat generation.

The discretization of the Pennes equation in the Cartesian grid is described in our previous work [14]. The discretized form in two dimensions is presented in the following formula:

$$T_{i,j}^{n+1} = T_{i,j}^n + \frac{\Delta t K}{\rho_{i,j} C_{i,j} \Delta x^2} \cdot \left[\begin{array}{c} T_{i-1,j}^n + T_{i+1,j}^n + T_{i,j-1}^n \\ + T_{i,j+1}^n - 4T_{i,j}^n \end{array} \right] \tag{2}$$

$$+ \frac{\Delta t}{\rho_{i,j} C_{i,j}} \left[(\omega_b)_{i,j} (\rho_b)_{i,j} (C_{Pb})_{i,j} \left(T_a^n - T_{i,j}^n \right) + Q_{i,j} \right]$$

The thermal properties used for temperature calculation of normal brain tissues and tumor are taken from our previous work [14]. A brain tumor with $\omega_B = 0.0016 S^{-1}$ and $Q_m = 25000 [\text{W/m}^3]$ is used in this work.

2.2 Level Set Method

Level sets method proposed by Osher and Sethian [8]. It is a powerful method widely used for brain tumor segmentation from MRI images. The main idea of the level set method is the implicit representation of the propagation of a curve C as a zero level set of higher dimensional level set function ($\varphi = 0$). The level set method evolves the curve by updating the level set function $\varphi(x, y, t)$ at fixed coordinates in the image over time. The level set function is a signed distance from each pixel over image to the curve and defined as follows:

$$\varphi(x, y, t = 0) = \pm d(x, y), \tag{3}$$

where, d is the signed distance function from (x, y) to C , this function Φ will evolve through time according to the speed term F defined by Sethian [15] via a partial differential equation:

$$\frac{\partial \varphi}{\partial t} = -|\nabla \varphi| \cdot F. \tag{4}$$

The evolution of the curve C in time is given by the zero level set with the following property:

$$C(t) = \{(x, y) | \varphi(x, y, t) = 0\}, \tag{5}$$

In Eq. (4) F is the speed function that describes the level set evolution. In this work, we adopt a simple speed function similar to [16, 17]; this speed function is a combination of two terms: curvature term for smoothness of the curve and data term for the curve propagation. Therefore, the level set equation takes the following form:

$$\frac{\partial \varphi}{\partial t} = -|\nabla \varphi| \cdot \left[\alpha D(I) + (1 - \alpha) \nabla \cdot \frac{\nabla \varphi}{|\nabla \varphi|} \right], \tag{6}$$

124   A. Bousselham et al.

where $\nabla \cdot (\nabla\varphi/|\nabla\varphi|)$ is the mean curvature term that keeps the level set function smooth during deformation, $D(I)$ represents the data function that causes the model to expand or contrast towards the segmentation sought in the input image data, and $\alpha \in [0, 1]$ is a free parameter used to control the smoothness of the curve.

The data term $D(I)$ forces the curve to grow in the desirable objects and contrast elsewhere in the input image. We have used a simple formulation of data function similar Lefohn et al. [16] and Cates et al. [17], it depends solely on the grayscale value of the input pixel intensity, which make $D(I)$ positive in tumorous regions or negative in healthy regions and described as follows:

$$D(I) = \varepsilon - |I - T|, \tag{7}$$

where, T represents the central intensity value of the region to segment, and ε is the intensity range around T that is part of the desired segmentation object sought.

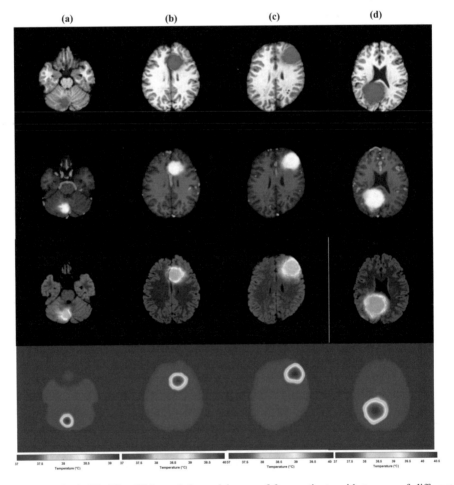

Fig. 1. Synthetic T1, T1c, Flair, and thermal image of four patients with tumors of different volumes. (a) Tumor with 11.6 cm^3 of volume. (b) Tumor with 27.4 cm^3 of volume. (c) Tumor with 51.1 cm^3 of volume. (d) Tumor with 81.7 cm^3 of volume.

Therefore when the grayscale intensity of the input pixel I is between $T - \varepsilon$ and $T + \varepsilon$, the curve expands and otherwise it contracts (Fig. 1).

2.3   Modified Speed Function

The main contribution of this work is to propose a new speed function, where we have changed the data term $D(I)$ by considering the temperature data instead of pixel intensity data:

$$D(\delta) = \varepsilon - |\delta - C|, \tag{8}$$

In the new data term, C represents the central temperature value of the region to segment, and ε is the temperature range around C that is part of the desired region, and δ is temperature data of the input pixel. Table 1 presents the used level set parameters values for the validation of the proposed approach, the threshold values are calculated using histogram where we have chosen the maximum pick in the tumorous region, and ε was chosen to have a maximum of accuracy.

Table 1. Level set parameters used for segmentation of brain tumors in four patients using different MRI images.

Sequence	Patient 1		Patient 2		Patient 3		Patient 4	
	T	ε	T	E	T	ε	T	E
T1	1016	220	1368	240	1498	240	1496	420
T1c	3983	1800	4007	1800	3932	1800	3516	1800
Flair	2516	480	2534	480	2577	550	2129	390
Thermal image	39	1.1	39	0.8	40	1.8	40	1.7

2.4   Synthetic MRI Images Used Experiments

The proposed approach is applied in synthetic MRI images of four patients with brain tumors taken from [18] with different volumes with ground truth, $11.6 \, cm^3$ for patient 1, $27.4 \, cm^3$ for patient 2, $51.1 \, cm^3$ for patient 3, and $81.7 \, cm^3$ for patient 4. The tumors were simulated using a software named TumorSim simulator [19]. This simulator combines physical and statistical modeling to generate synthetic multi-modal 3D of T1, T2, FLAIR, and post gadolinium T1 contrast MRI volumes with realistic tumors. It was widely employed to evaluate brain tumor segmentations in recent works [20, 21]. The provided database contains 100 synthetic images with varying locations and volumes of tumors created from 20 patients in the BrainWeb database five images per patient [22].

The level set method and Pennes equation where implemented using C language on Windows 7 operating system with a CPU Intel i7-4770k. The C code has been compiled with Visual C++ compiler. ITK (www.itk.org) library was used to read and save DICOM images.

3   Results and Discussion

In the following, we present the obtained results of segmentation using level set method applied in T1 and T1c MRI images, and the modified level set method in the calculated thermal image. Figure 1 presents the used images for validation of the proposed approach, which contains 2D images of synthetic T1, T1c, Flair of four patients and the calculated thermal images using Pennes bioheat equation. The 2D slices are taken from 3D MRI volume where the tumor surface is maximum.

Figure 2 presents the obtained results of the segmentation. Each patient is shown in a different column; the first three lines present the segmentation results using level set method applied in MRI T1, T1c, and Flair sequences. The last line gives the obtained results of segmentation using level set method applied in the thermal image; the results

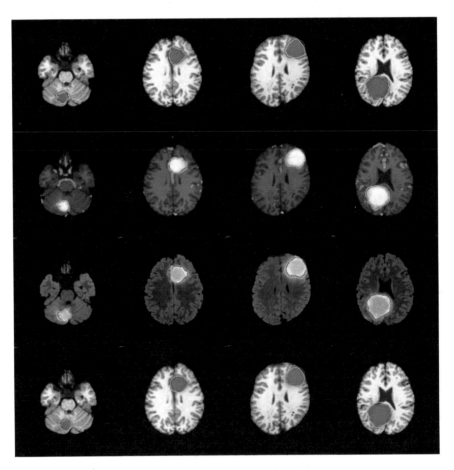

Fig. 2. Results of segmentation by level set method in MRI images and in thermal image. The first, second, and the third lines provide the segmentation by level set in T1, T1c, and Flair. The last line gives the segmentation by level set method in thermal image. Green: segmentation, Red: ground truth.

of segmentation are overlaid in T1-weighted images. At first glance, we can observe that there is a significant improvement using level set method in the thermal image as the temperature showed a high variation in tumor borders. Also, the obtained curve in thermal image is smooth compared to the conventional MRI sequences; this can be explained as the used model of bioheat equation is isotropic, therefore in future works, we plan to use anisotropic model for more accurate temperature estimation.

Table 2 shows the percent of the tumor segmented using level set method in thermal image and segmented using level set method in T1, T1c, and Flair sequences. The obtained improvement is 10.29 % for T1, 4.32 % for T1c, 22.58 % for Flair. These results prove the importance of temperature in the diagnostic of brain tumors, which gives additional information about its borders that can be used to improve and reinforce the accuracy of segmentation algorithms in MRI.

Table 2. The percent of tumor pixels segmented by level set in thermal image, not segmented by level set in T1, T1c, and Flair for four patients.

	Patient 1	Patient 2	Patient 3	Patient 4
T1	16.69	10.68	10.67	3.13
T1c	7.92	1.63	1.6	6.13
Flair	17.36	32.15	0.92	39.9

In the present study, we used level set method for brain tumor segmentation from temperature distribution (thermal image). Tumors show a high temperature compared to healthy tissues due to heat generation by cells metabolism. Temperature reveals a high variation in tumor contours. Thus, we have applied level set method to locate and estimate the tumor margins. The obtained results showed that segmentation in the thermal image could be used to improve and reinforce segmentation methods in conventional MRI sequences (T1, T1c, and Flair). To the best of our knowledge, we are the first to incorporated thermal analysis to improve the accuracy of brain tumor segmentation in MRI.

4   Conclusion

In this study, we presented a new approach for brain tumor segmentation using a level set method based on temperature changes in the pathologic area. The thermal images are calculated using Pennes bioheat equation, and then level set method was applied to localize tumor contours. The proposed approach proved that temperature is a good way that can be used to reinforce the conventional MRI sequences such as T1, T1c and Flair sequences. As a future work, we plan to modify the Pennes equation to make it anisotropic to calculate the temperature with more accuracy, as the biological tissues are anisotropic. Also, we will apply other segmentation methods on the thermal image to compare the obtained results with the results in the present paper and apply them in realistic images.

Acknowledgements. This work is supported by the grant of the National Center for Scientific and Technical Research (CNRST - Morocco) (No. 13UH22016).

References

1. Helms, G., Kallenberg, K., Dechent, P.: Contrast-driven approach to intracranial segmentation using a combination of T2-and T1-weighted 3D MRI data sets. J. Magn. Reson. Imaging **24**, 790–795 (2006)
2. Jin, L., Min, L., Jianxin, W., Fangxiang, W., Tianming, L., Yi, P.: A survey of MRI based brain tumor segmentation methods. Tsinghua Sci. Technol. **19**(6), 578–595 (2014)
3. Bauer, S., Wiest, R., Nolte, L.P., Reyes, M.: A survey of MRI based medical image analysis for brain tumor studies. Phys. Med. Biol. **58**(13), 97–129 (2013)
4. Gordillo, N., Montseny, E., Sobrevilla, P.: State of the art survey on MRI brain tumor segmentation. Magn. Reson. Imaging **31**(8), 1426–1438 (2013)
5. Dvorak, P., Menze, B.: Structured prediction with convolutional neural networks for multimodal brain tumor segmentation. In: Proceedings MICCAI BraTS (Brain Tumor Segmentation Challenge), pp. 13–24 (2015)
6. Havaei, M., Dutil, F., Pal, C., Larochelle, H., Jodoin P.-M.: A convolutional neural network approach to brain tumor segmentation. In: Proceedings MICCAI BraTS (Brain Tumor Segmentation Challenge), pp. 29–33 (2015)
7. Kass, M., Witkin, A., Terzopoulos, D.: Snakes: active contour models. Int. J. Comput. Vis. **1**(4), 321–331 (1988)
8. Osher, S., Sethian, J.A.: Fronts propagating with curvature dependent speed: algorithms based on hamilton-jacobi formulations. J. Comput. Phys. **79**(1), 12–49 (1988)
9. Bousselham, A., Bouattane, O., Youssfi, M., Raihani, A.: 3D brain tumor localization and parameter estimation using thermographic approach on GPU. J. Therm. Biol. **71**, 52–61 (2018)
10. Sadeghi-Goughari, M., Mojra, A.: Finite element modeling of haptic thermography: a novel approach for brain tumor detection during minimally invasive neurosurgery. J. Therm. Biol. **53**, 53–65 (2015)
11. Sadeghi-Goughari, M., Mojra, A.: Intraoperative thermal imaging of brain tumors using a haptic-thermal robot with application in minimally invasive neurosurgery. Appl. Therm. Eng. **91**, 600–610 (2015)
12. Pennes, H.H.: Analysis of tissue and arterial blood temperatures in the resting human forearm. J. Appl. Physiol. **1**(2), 93–122 (1948)
13. Wissler, E.H.: Pennes' 1948 paper revisited. J. Appl. Physiol. **85**(1), 35–41 (1998)
14. Bousselham, A., Bouattane, O., Youssfi, M., Raihani, A.: Thermal effect analysis of brain tumor on simulated T1-weighted MRI images. In: International Conference on Intelligent Systems and Computer Vision (ISCV), April 2018
15. Sethian, J.A.: Level Set Methods and Fast Marching Methods. Cambridge University Press, Cambridge (1996)
16. Lefohn, E.A., Kniss, M.J., Hansen, D.C., Whitaker, T.R.: A streaming narrow-band algorithm: interactive computation and visualization of level sets. IEEE Trans. Vis. Comput. Graphics. **10**, 422–433 (2004)
17. Cates, J.E., Lefohn, A.E., Whitaker, R.T.: GIST: An interactive, GPU-based level set segmentation tool for 3D medical images. Med. Image Anal. **8**, 217–231 (2004)

18. Galimzianova, A., Pernus, F., Likar, B., Spiclin, Z.: Robust estimation of unbalanced mixture models on samples with outliers. IEEE Trans. Pattern Anal. Mach. Intell. **37**(11), 2273–2285 (2015)
19. Prastawa, M., Bullitt, E., Gerig, G.: Simulation of brain tumors in MR images for evaluation of segmentation efficacy. Med. Image Anal. **13**(2), 297–311 (2009)
20. Ahlgren, A., Wirestam, R., Ståhlberg, F., Knutsson, L.: Automatic brain segmentation using fractional signal modeling of a multiple flip angle, spoiled gradient-recalled echo acquisition. Magn. Reson. Mater. Phys. **27**, 551–565 (2014)
21. Nabizadeh, N., John, N., Wright, C.: Histogram-based gravitational optimization algorithm on single MR modality for automatic brain lesion detection and segmentation. Expert Syst. Appl. **41**, 7820–7836 (2014)
22. Aubert-Broche, B., Griffin, M., Pike, G.B., Evans, A.C., Collins, D.L.: Twenty new digital brain phantoms for creation of validation image data bases. IEEE Trans. Med. Imaging **25** (11), 1410–1416 (2006)

Context-Awareness-Based Adaptive Modal for Multimedia Documents

Hajar Khallouki[1(✉)], Khaoula Addakiri[2], and Mohamed Bahaj[1]

[1] Mathematics and Computer Science Department, Hassan I University,
Faculty of Sciences and Techniques, Settat, Morocco
hajar.khallouki@gmail.com
[2] Computer Science Department, Ibn Zohr University, Ouarzazate, Morocco

Abstract. Pervasive computing is the result of the convergence of the field of mobile computing and distributed systems. Pervasive computing is based on knowing the user's context in order to provide a service adapted to its context that is further exploited by multimedia document adaptation processes. In this paper, we introduce an approach for modeling context in pervasive computing environments for multimedia documents adaptation. The main benefit of this model is the ability to reason about various contexts and generate pertinent adaptation services.

Keywords: Pervasive computing · User's context · Multimedia document ·
Adaptation services

1 Introduction

Computer systems have grown considerably in recent years. This development is marked by the technological evolution of mobile terminals (use of smart cell phones, personal digital assistants (PDAs), etc.).

The evolution of mobile technology is also reflected in the development of wireless communications networks (Wi-Fi, Wimax, Bluetooth, etc.). All of these technologies completely change the way computers are used. With the advent of these mobile technologies, we are talking about pervasive computing, ubiquitous or ubiquitous systems. Ubiquitous Computing makes the use of Computing possible at any time and regardless of the user's position. This new form of Information Technologies (IT) opens up new opportunities for users to access shared and ubiquitous services centered on their needs.

In this paper, we begin by presenting some basic concepts in this area and notions that we come across throughout this document. We focus on the concept of context and the principle of context-awareness which constitute the central notions of our research work. Thus, we introduce our proposed approach which is based on context-awareness for multimedia documents adaptation. Finally, we conclude the paper and discuss future work.

© Springer Nature Switzerland AG 2019
F. Khoukhi et al. (Eds.): AIT2S 2018, LNNS 66, pp. 130–134, 2019.
https://doi.org/10.1007/978-3-030-11914-0_14

2 Related Work

The development of communication tools also facilitated the exchange between different computer equipment to cooperate in a fluid and fast manner. This trend towards the systematic computerization of resources promoting access and exchange of information anywhere and at any time is qualified by pervasive computing [1].

Schilit and Theimer [2] defined the context as the changes in the physical environment according to: location, description of people and objects in the environment and the changes of these objects.

The context cannot be taken independently of the spacio-temporal constraint. This is in line with the opinion of Ryan et al. [3] on the importance of time in the characterization of the context. Indeed, any situation, event or interaction is necessarily fulfilled in a time interval. The granularity of this interval is crucial since it specifies the temporal extent of the information describing the context. This fact also influences the amount of information: we can go from a small amount of information to a database, depending on whether we are interested in an isolated event (short-term event) or in a sequence of events.

In the literature, the spatial context is reduced to the localization variable (location of the user) in the best case. We can identify places in the user's environment [4]. Rather, we can find the notion of "context of use" which defines and encompasses all information characterizing the user, the environment and the platform [5].

The interaction in a pervasive system has to be defined according to two axes: (1) of human-machine interaction exposed by Dey in his definition of the context, (2) of machine-machine interaction. According to our study, this last axis does not seem to have been taken into consideration. However, this aspect is becoming increasingly important given the increasing development of M2M (machine to machine) infrastructures and web services where the contextual information relating to the interaction between applications can support the responsiveness and efficiency of these systems to the respect of the user. This aspect is defended by Brézillon and Coutaz who consider "the context as an interactional problem in which the interaction itself can be important" [6].

Bucur et al. [7] have deepened the characterization and formulation of the relevance of contextual information by integrating the concept of purpose into context (defining the contextual elements according to the purpose of their use). Purpose refers to the purpose for which the context is used at a given point in time. This goal can be short or long term.

Some researchers have proposed approaches for modeling the quality and relevance of contextual information. For example, Razzaque, Dobson, and Nixon [8] have developed a generic approach that consists of a stepwise process. While Krause and Hochstatter, [9] have proposed a method for representing integrated context quality aspects in their context meta-model (CMM) representation model.

3   Contribution

Our architecture aims to make the multimedia documents adaptation more generic using multimodal interactions. The user can display the multimedia documents easily according to his context and preferences. The architecture contains different layers (see the Fig. 1).

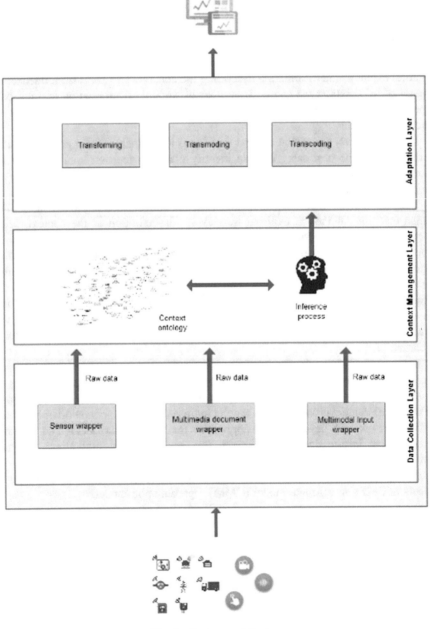

Fig. 1. System architecture

Context platform can be used to sense, process, store, and transfer context data. A Sensor provides an API to obtain the captured data and to configure the notifications (for example to send the measurements every three seconds or to signal a threshold overrun). It represents a physical sensor (for example a GPS location sensor) or a logic sensor (for example a software component for measuring the CPU load).

The first step of enabling smart services is to collect contextual information about environment, location, user input, etc. For example, sensors can be used to continuously monitor human's physiological activities and actions such as health status and motion patterns.

Data Collection Layer

The data collection layer collects and processes multimedia document characteristics such as codec and resolution, multimodal inputs such as speech and hand gesture, as well as sensed data such as location and temperature.

The sensor wrapper collects raw context data using various sources, such as sensors, and processes them into context markup. In this, the sensor wrapper automatically transforms the received sensed data into context markup. The multimodal wrapper collects user's preferences expressed through their modalities, The multimodal services allow the recognition of the input modalities and present their semantic. There are many multimodal services integrated in devices such as; the HandWriting Recognition (HWR)1, Speech Recognition (SR)2 and EmotionML3 for user emotions recognition. Finally, the multimedia document wrapper collects the received multimedia objects characteristics.

Context Management Layer

The context management layer is composed of two components; the context onlogy which presents a semantic representation of user context through a model that allows users to control how to receive their multimedia documents in a given condition.

The reasoning process allows generating a set of actions that will be executed by the adaptation services.

Adaptation Layer

The adaptation layer allows the transformation of a multimedia object into another multimedia object satisfying a given context. Two types of adaptations can be distinguished.

- Technical adaptation: This adaptation is related to the capacity of mobile devices (memory, display, etc.). It has two parts:
  - Transcoding: conversion of format, e.g., JPEG to PNG.
  - Transmoding: conversion of types, e.g., text to sound.
- The semantic adaptation: this adaptation is related to the constraints of the data types handled by mobile devices. It allows the content change without changing the media type and format, e.g., text summarization, language translation, etc.

4 Conclusion

This paper introduces context-aware architecture based on different components which enable the user context prediction and generate a pertinent adaptation services.

As discussed above, there remains more work to be carried out for covering all the aspects of multimedia documents and for deepening the specification of documents and adaptation conditions so that the adaptation produces quality results.

References

1. Krupitzer, C., Roth, F.M., VanSyckel, S., Schiele, G., Becker, C.: A survey on engineering approaches for self-adaptive systems. Pervasive Mob. Comput. **17**, 184–206 (2015)
2. Schilit, B.N., Theimer, M.M.: Disseminating active map information to mobile hosts. IEEE Network **8**(5), 22–32 (1994)
3. Baldauf, M., Dustdar, S., Rosenberg, F.: A survey on context-aware systems. Int. J. Ad Hoc Ubiquitous Comput. **2**(4), 263–277 (2007)
4. Kapitsaki, G.M., Prezerakos, G.N., Tselikas, N.D., Venieris, I.S.: Context-aware service engineering: a survey. J. Syst. Softw. **82**(8), 1285–1297 (2009)
5. Najar, S., Pinheiro, M.K., Souveyet, C.: A new approach for service discovery and prediction on pervasive information system. Procedia Comput. Sci. **32**, 421–428 (2014)
6. Gensel, J., Villanova-Oliver, M., Kirsch-Pinheiro, M.: Modèles de contexte pour l'adaptation à l'utilisateur dans des Systèmes d'Information Web collaboratifs. In: Workshop from" 8èmes journées francophones". Sophia-Antipolis, France (2008)
7. Bucur, O., Beaune, P. Boissier, O.: Définition et représentation du contexte pour des agents sensibles au contexte. In: Proceedings of the 2nd French-speaking Conference on Mobility and Ubiquity Computing, ACM (2005)
8. Razzaque, M.A., Dobson, S., Nixon, P.: Categorization and modelling of quality in context information (2006)
9. Fuchs, F., Hochstatter, I., Krause, M., Berger, M.: A metamodel approach to context information. In: PerCom Workshops (2005)

E-learning

LMOnto: An Ontology-Based Learner Model for Technology Enhanced Learning Systems

Laila Akharraz[1]([✉]), Ali El Mezouary[1], and Zouhir Mahani[2]

[1] IRF-SIC Laboratory, IBN ZOHR University, Agadir, Morocco
Laila.akharraz@edu.uiz.ac.ma, a.elmezouary@uiz.ac.ma
[2] LASIME Laboratory, IBN ZOHR University, Agadir, Morocco
z.mahani@uiz.ac.ma

Abstract. Several initiatives concerning the learner model has been proposed in the context of technology-enhanced learning systems. In fact, there are two main challenges encountered in this regard: Firstly, which information will serve to represent and reflect better the learner? Secondly, which formalism chooses for representing and managing the learner model? To overcome those challenges, we propose a new ontological approach for learner modeling enriched from the existing models.

Keywords: Learner modeling · Learner model · Ontology ·
Technology enhanced learning system · Learning context

1 Introduction

In technology enhanced learning system, learners are generally heterogeneous. They have different needs, different intellectual abilities, different learning styles and different goals and preferences. Thus, learners learn differently and if we use the same courses and the same learning styles in all situations, the quality of understanding will be less for most learners. Moreover, the learner's knowledge evolves over time and since learners have diverse and varied attitudes. All these characteristics need to be modeled to adapt learning environments to the learner's needs.

The learner model is a representation of knowledge and information about a learner (i.e. knowledge, backgrounds, skills, problem-solving strategies, preferences, performances, learning styles, goals) [1]. A literature review of [2–4] was conducted to identify the fundamental roles of learner modeling that may be listed as follows:

- Detecting, preventing and correcting errors committed by the learner.
- Determining the next appropriate content that will be presented to the learner.
- Evaluating the learner.
- Predicting the future behavior of learners.
- Monitoring the effectiveness of the learning strategy used.
- Following up the learner and predicting his future behaviors.

© Springer Nature Switzerland AG 2019
F. Khoukhi et al. (Eds.): AIT2S 2018, LNNS 66, pp. 137–142, 2019.
https://doi.org/10.1007/978-3-030-11914-0_15

Moreover, the learner model allows the system to react in accordance with the learner's characteristics and its interactions with the system [5]. The modeling process requires a set of treatment to collect and update relevant information about the learner [6].

For a better representation of the learner, we thought about using ontology. In this paper, we propose a learner model ontology (LMOnto) including the different characteristics of a learner.

2 Related Work

There are several approaches to construct the learner model in Technology Enhanced Learning systems: the stereotype model in which the learners are classified into distinct groups or categories according to certain characteristics that they shared [7], the overly model [8] in which the learner's knowledge is only a subset of the expert's knowledge; the differential model (overlay model extended) in which the learner's knowledge is classified into two categories: The knowledge that the learner is supposed to acquire and the knowledge not yet presented to the learner [9]; the perturbation model (BUGGY model) developed by [10] which incorporates false knowledge, corresponding to erroneous knowledge of the learner. Therefore, there are other important technologies for learner model representation, such as: the machine learning techniques allow to observe automatically the learner's actions, then to extract automatically information about the learner, the constrained-based model proposed by [11] where the domain knowledge is represented as a set of constraints and the learner model is the set of constraints that have been violated, the Bayesian networks for representing and dealing with the uncertainty in the learner model where every node represents the different components/dimensions of a learner such as knowledge, Learning Styles, emotions, goals, etc. [12], the fuzzy logic modeling techniques which is also able to deal uncertainty in learner model due to imprecise and incomplete data, and ontological technology which has been successfully used for learner modeling [13–15]. The ontology was used for tow mean reasons, "(a) ontology supports the formal representation of abstract concepts and properties in a way that they can be reused by many tasks or extended if needed and (b) they enable the extraction of new knowledge by applying inference mechanisms (e.g. reasoner) on the information presented in the ontology" [16].

Furthermore, a number of researchers proposed a hybrid learner model based on combining one or more of the above approaches conforming to needs and purposes of the system.

3 LMOnto: The Learner Model Ontology

The first stage of building a learner model is the selection of pertinent learner characteristics that should be taken into consideration. These characteristics may differ depending on modeling's aims adaptation are reviewed. The objective of this work is to propose a new approach of ontology-based learner modeling to simulate the learning context, the environment and the behavior of the learner in order to better evolve the learner model. In this section, we describe the content of proposed learner model.

3.1 Required Information for Building Learner Model

The proposed learner model is defined as ontology, including the different characteristics of a learner. We propose to describe a learner according to four main categories of information.

- Personal Data: describes the learner's personal data including general information such as name, surname, age, gender, learning style, goals, preferences, and the password and the learner's login to access the environment.
- Cognitive Data: concerns the general cognitive characteristics of the learner such as his prerequisites, his linguistic skills, his general technical skills, as well as his competences concerning specific disciplines according to the field of learning.
- Contextual Data: integrate all the data relating to the learner context and which can have an influence on its learning, among these data, we quote: the learning device describing the tools and the equipment's that the learner uses (computer, mobile, tablet), the screen resolution, the browser used by the learner, in addition to the information about the learners accessibility to the learning system (access dates, duration of each session, the frequency of access, etc.).
- Activity Data: includes all information relating to the assessment and history of the learner's activities in the system, regardless of the type of activity: courses, tests, quizzes or "practical work" activities. The learner information can be obtained in different ways: so as an explicit acquisition when information is given manually (i.e. forms, QCMs), or an implicit acquisition which allows retrieving information automatically from the environment, and acquisition by inference from the analysis of interaction between the learner and the environment.

3.2 Description of LMOnto

In order to construct the model described in the previous section, we adopted the Web ontology language (OWL). The development process of the ontology was accomplished with the aid of Protg3 tool.

The information represented by an ontology is characterized by five elements: concepts, relations, functions, Axioms and instances. In this section, we describe the elements implemented in our learner model ontology. As it presented in Fig. 1, LMOnto has four main classes: PersonalData, CognitiveData, ActivityData, and Context.

Table 1 presents a brief description of ActivityData class by specifying the super Class, the properties, and the relationships.

Table 1. Description of the class ActivityData

Concept	Super class	Object properties	Data properties	Description
Learner Activity	ActivtyData	hasDuration began hasRessourType hasLog	Activity_id Activity_description Activity_begin Activity_Type Activity_Objective	Activities done by the learner including current activity and finished activity
Evaluation Log	EvaluationLog	isAssociatedWith (activity)	-	Log for keeping track the history of learner activities
Errors	EvaluationLog	isAssociatedWith (activity)	Error_frequency Error_type	Mistakes made by the learner during the educational process
OutCome	EvaluationLog	isAssociatedWith (activity)	-	Learning outcomes succeeded
Knowledge Gained	EvaluationLog	isAssociatedWith (activity)	-	Knowledge obtained by the learner during the educational process

Furthermore, the proposed learner model ontology has been enriched with a set of rules in order to enable the extraction automatically of new knowledge by applying inference mechanisms (e.g. reasoner) on the information presented in the ontology. All the rules are expressed in Semantic Web Rule Language4 (SWRL). For instance, in Table 2 presents the following rule: "if the learner has finished an activity which required a competency as prerequisites then the learner get this competency".

Table 2. Example of SWRL rules

	Rule body	The result
If	A isA LearnerActivity	**Then** L hasCompetency C
And	C isA Cometency	
And	A hasPrerequisite C	
And	L isA Learner	
And	L hasFinichedActivity A	

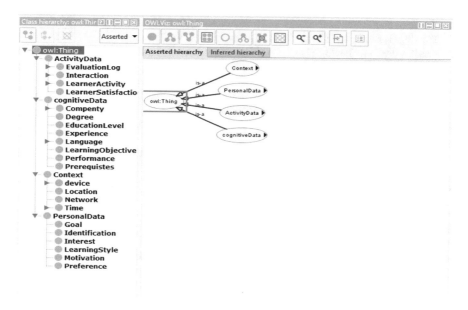

Fig. 1. The LMOnto ontology as displayed in Protégé

4 Conclusion and Perspectives

Learner modeling is a very active field of research in the technology enhanced learning system community. There are a tree constraints related to this topic. In order to handle them we must answer for those questions: (a) what are relevant characteristics to represent in the learner model? (b) What is the most appropriate formalism to represent the learner? (c) What is the modeling process to adopt in order to better evolve the learner model? In this paper, we answered the first and second question by presenting a new approach of ontology-based learner modeling to simulate the learning context, the environment and the behavior of the learner. Such modeling will allow obtaining as a result a robust learner model which leads to an adaptation of the learning contents consequently to the improvement of the learning situations. The future work consists of proposing a prototype that will be integrated into Moodle platform in order to exploit efficiently the learner model ontology developed.

References

1. Chrysafiadi, K., Virvou, M.: Student modeling approaches: a literature review for the last decade. Expert Syst. Appl. **40**(11), 4715–4729 (2013)
2. Self, J.: The role of student models in learning environments. IEICE Trans. Inf. Syst. **77**, 3–8 (1994). No. 94
3. Jameson, A., Paris, C., Tasso, C.: User Modeling: Proceedings of the Sixth International Conference, UM97. SpringerWein, New York (1997)

4. VanLehn, K.: Toward a theory of impasse-driven learning. In: Learning Issues for Intelligent Tutoring Systems, pp. 19–41. Springer, New York (1988)
5. Brusilovsky, P., Peylo, C.: Adaptive and intelligent web-based educational systems (IJAIED). Int. J. Artif. Intell. Educ. **13**, 159–172 (2003)
6. Mccalla, G., Vassileva, J., Greer, J., Bull, S.: Active learner modelling (2000)
7. Kay, J.: Stereotypes, student models and scrutability. In: International Conference on Intelligent Tutoring Systems, pp. 19–30 (2000)
8. Carr, B., Goldstein, I.P.: Overlays: a theory of modelling for computer aided instruction (1977)
9. Greer, J.E., McCalla, G.I.: Student Modelling: The Key to Individualized Knowledge-Based Instruction, vol. 125. Springer, Heidelberg (2013)
10. Brown, J.S., VanLehn, K.: Repair theory: a generative theory of bugs in procedural skills. Cognit. Sci. **4**(4), 379–426 (1980)
11. Ohlsson, S.: Constraint-based student modelling. J. Interact. Learn. Res. **3**(4), 429 (1992)
12. Zhouand, X., Conati, C.: Inferring user goals from personality and behavior in a causal model of user affect, pp. 1–8 (2003)
13. Panagiotopoulos, I., Kalou, A.: An ontology-based model for student representation in intelligent tutoring systems for distance learning. In: Artificial Intelligence Tutoring Systems for Distance Learning, pp. 296–305 (2012)
14. Rezgui, K., Mhiri, H., Ghédira, K.: An ontology-based profile for learner representation in learning networks. Int. J. Emerg. Technol. Learn. **9**(3), 16–25 (2014)
15. Kikiras, P., Tsetsos, V., Hadjiefthymiades, S.: Ontology-based user modeling for pedestrian navigation systems. In: ECAI 2006 Workshop on Ubiquitous User Modeling (UbiqUM), Riva del Garda, Italy, pp. 1–6, January 2006
16. Clemente, J., Ramírez, J., De Antonio, A.: A proposal for student modeling based on ontologies and diagnosis rules. Expert Syst. Appl. **38**(7), 8066–8078 (2011)

Context-Aware that Performs Adaptation in Mobile Learning

Abdelkarim Aziz$^{(\boxtimes)}$ and Faddoul Khoukhi$^{(\boxtimes)}$

Faculty of Science and Technics, University Hassan II Casablanca,
Mohammedia, Morocco
`karim.aziz@karimoosteam.com, khokhif@gmail.com`

Abstract. Mobile computing is recognized as the new emerging technology to enhance teaching approaches and provide new learning strategies that exploit learners' contextual information. The context aware in mobile learning could load dynamically information's form learners and exploit that information's to provide a personalized user experience according to it context. The main challenge in context-aware M-learning system is context ontologies, it represents the most part for context modeling. In this paper, we propose an ontology-based approach for modeling context-aware in an adaptive mobile learning environment.

Keywords: Mobile learning · Adaptive learning · Ontology engineering · Context awareness · Software engineering

1 Introduction

With the rapid progress of mobile computing technology and network connections, mobile learning has gained significant potential in the field of e-learning and the new teaching approaches [1], Learners are in the center of the learning process and they are provided with adaptive learning experience related to their personal characteristics and educational objectives. In general adaptation can be done based on learner profile, learning content, in this paper we explore the adaptation that can be done on the learning environment. In order to provide adapted services for learner, applications and services should be aware of their contexts and automatically adapt to their changing context-known as context awareness. Context is any information that can be used to characterize the situation of an entity. An entity is a person, place, or object that is considered relevant to the interaction between a learner and an application, including location, time, activities, and the preferences of each entity [2]. Context awareness is increasingly taking applicability in pervasive mobile computing, this paradigm refers to the idea that the context aware mobile learning system should be able to capture, analyze and manage every state of learner, his external environment and adapt it reactions to the current situation needs [3].

Using ontology to model the context aware in mobile learning environment has already been proposed in many approaches, but most only offer a weak support for adaptation services, knowledge sharing, context reasoning and context information presenting. Ontology modeling has proved an effective way to deal with context

© Springer Nature Switzerland AG 2019
F. Khoukhi et al. (Eds.): AIT2S 2018, LNNS 66, pp. 143–152, 2019.
https://doi.org/10.1007/978-3-030-11914-0_16

awareness in ubiquitous learning environment for the following reasons [4]. 1. Ontologies are powerful tool to deal with knowledge sharing and information reuse. 2. Ontologies enable context-reasoning and interoperability in a pervasive computing system. 3. Ontologies play a major role in semantic description modeling of information. Ontology represents a description of concepts and their relationship; these models are very promising for modeling contextual information due their high and formal expressiveness and possibilities for applying ontology reasoning technics [5]. Ontologies in education can be used in various field, they include domain modeling [6], context modeling and reasoning [7], learning object creation [8] and semantic context aware reasoning [9].

In this study, we propose an ontology-based approach for modeling learner contextual information in order to provide learners with appropriate content related to their context of learning. This paper is structured as follows: In Sect. 2 we explore the notion of context awareness and his relation with adaptability; in Sect. 3 we introduce our context aware modeling based on ontology and semantic web and we conclude in the last section.

2   Related Work

The most relevant approaches presented in the literature for modeling a context aware are [10–12]:

2.1   Key Value Models

In this model context information are presented as key-value link process, the key presents the information identity and value present the current value of this information (e.g. Set location = room_1). The advantage of this model is the facility of implementation however it presents an inconvenient of information expressiveness for more sophisticated structuring purpose.

2.2   Graphic Model

This model uses unified modeling language (UML) to model the context. In particularly this model is useful for structuring context information, but usually not used on instance level.

2.3   Markup Scheme Model

This model is an XML based model, it uses markup query language method to retrieve context information and it defined as an extension of composite capabilities (preference-profile) [13], user agent profile (UA prof) encoded in RDF.

2.4   Object Oriented Model

The main intention behind object orientation model is encapsulation and reusability that offer oriented object programming.

2.5   Logic Based Model

This model provides a high level of formality, the context in this model is defined as facts, expression and rules, however it presents an inconvenient such as dealing with uncertain or conflicting information.

2.6   Ontology Based Model

Ontologies are very powerful tools for modeling context in mobile environment; the context is modeled as concept and fact based on reasoning methods for retrieving context information. It presents a formal model of context that can be shared, reused, extended and combined. As an example; SOCAM [14] (Service Oriented Context-Aware Middleware), COBRA [15] (Context Broken Architecture) are context aware middleware's that provide specifications and requirements to develop a context aware system but all offer a weak support of general adaptation service.

The advantage of ontologies resides in the semantic description of context information and reasoning technics engines. According to [16] context modeling should be satisfied five requirements:

- Distributed Composition (DC)

Mobile learning systems are a distributed system, with no central instance responsible of the creation, deployment and maintenance of information.

- Partial Validation (PV)

In a context modeling approach, it's highly recommended to partially validate contextual knowledge on structure as well as on instance level (against a context model in use) because of the complexity of the contextual information relationship.

- Richness and Quality of Information (RQI)

The quality of contextual information captured by sensors varies over time, so, the context modeling should be aware of that and take advantage of it.

- Incompleteness and Ambiguity (IA)

The model should be aware of uncompleted information captured (e.g. Interpolated instance)

- Applicability to Existing Environment (AEE)

The context model should be implemented and deployed within an existing platform.

The above-mentioned requirements are important to model a context in ubiquitous environment such M-learning; however, we find that another requirement should be added to this list in order to deal with learning adaptation.

- Context Adaptation (CA)

The context-aware model should be aware of learner contextual information, and use them as input data to provide adapted content to leaner profile.

The Key value model validate one of the requirements, XML-model, graphic model, logic-based model validate more than one. Ontology and oriented object model validate all the requirements, only the ontology provides a better understanding of context information and a better validation of partial validation.

We summary the entire requirements in the Table 1 as follow:

Table 1. Context aware requirements pattern using ontology [5, 17].

Approach	Requirements					
	DC	PV	RQI	IA	AEE	CA
Key-value model	–	–	–	–	–	–
Mark-up scheme models	+	++	–	–	+	–
Graphical models	–	–	+	–	+	–
Oriented object models	++	+	+	+	+	–
Logic based models	++	–	–	–	++	–
Ontology based models	++	++	+	+	++	++

3 Context-Aware that Performs Adaptation in Mobile Learning

The future of software engineering todays is gradually converted from "software-centric" to "Learner-centric" named Learner experience. It's primary important to provide learners with accurate, efficient and adapted information related to their context of use as context aware which result in a great learner experience with the software.

Context is any information that can be used to characterize the situation of an entity, an entity can be a person, location, time, device or object that is relevant to the interaction between a learner and a software [2]. Adaptation through context play a major role in the process of learning experience, it provides the appropriate content for learner based on his contextual information. In addition, applications and services should be aware of their context and automatically adapt their behaviors to the current detected context as context-awareness.

In mobile learning a context aware can provide a significant adaptation of their services to the learners needs, it should be able to detect, interpret and use context information such as location, time, and any information that can be used to improve an adapted learning process.

4 Toward an Ontology-Based Approach of Context Aware

The learning context is a critical aspect in the field of mobile leaning. In recent years a new focus has been formulated, the learning process is changing the way learners learn and progress, the new approach focuses on the creation of mobile learning environment that support adaptive learning as well provide personalized learning scenario to learner profile based on his context of use. Context is one of the key issues for adaptive learning; in this paper, we propose a new approach for modelling a context aware adaptive learning environment in order to meet learner needs. This approach is based on five elements as follow:

- Spatial Context awareness
- Temporal Context awareness
- Learner Context awareness
- Device Context awareness
- Vector Context awareness

All elements are briefly detailed in the next section.

4.1 Spatial Context Awareness

We find that spatial location of learners is an important factor in order to provide content related to his location, we aim to propose a metadata that defines the spatial context.

This metadata is based on the following attributes:

_Location Name: it presents the location name (e.g. FST_LIM).

_Location Type: it presents the location type (e.g. LIM is a laboratory of computer science in FST).

_Location Properties: it presents the properties related to learner location, in some cases the context aware environment should be able to detect and manage the characteristics of the environment. (e.g. location with a high level of noise, the system should manage the volume level of the device).

_Geographic Position (Latitude, Longitude): we attempt to use the global spatial database (e.g. GeoNames server) to collect name, address and type of location in use (e.g. learner is located in LIM laboratory, so we provide him recent content related to computer science).

_Concept Interrelationship: ontology and semantic web provide a variation of semantic relationship to determine constraints between concepts presented above.

We present spatial context awareness as "SpatialContext", the Fig. 1 shows a brief taxonomy of concept and their semantics relationship.

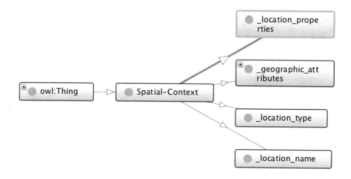

Fig. 1. Spatial context ontology

4.2   Temporal Context Awareness

Also, we find temporal context of learner an important factor in the adaptation mechanism. ISO 8601 [18] provides a full standard way to write date, time of day, UTC, Time, etc. In this case, not just we adapt learning content to learner location but also, we provide him options to select his appropriate learning timing. The recent OWL Time ontology provides full representation of time [19].

We aim to propose a metadata that defines the temporal context; this metadata is based on the following attributes:

- *TimeOfDay*: it presents periodic timing of learner (e.g. Night).
- *DayOfWeek*: it presents the day of the week (e.g. Sunday).
- *MonthOfYear*: it presents the month of the year (e.g. April).
- *FreeTime*: it presents the free time of the user.
- *SeasonOfYear*: it presents the season of the current year.
- Date-Time: it presents a description of date and time structured with separate values for the various elements of a calendar-clock system, the temporal reference system is fixed to Gregorian Calendar, and the range of year, month, day properties restricted to corresponding XML Schema types xsd:gYear, xsd:gMonth and xsd: gDay, respectively.
- *Clock*: this class the current clock system e.g. learner start learning at 10:00 pm.
- *TimeZone*: A Time Zone specifies the amount by which the local time is offset from UTC. A time zone is usually denoted geographically (e.g. Australian Eastern Daylight Time), with a constant value in a given region. The region where it applies and the offset from UTC are specified by a locally recognized governing authority.
- *TimePosition*: A temporal position described using either a (nominal) value from an ordinal reference system, or a (numeric) value in a temporal coordinate system.
- Laps: it presents the interval timing of learning related to the context of learner (e.g. Learner in bus has 30 min of road then Laps = 30 min), we use relations provided by Allen and Ferguson [20] to represent a laps of learning time. (E.g. StartTime indicate the start of the learning process).

We present Temporal Context awareness as temporal Context; therefore, this context has been described into three categories: Abstract Time (Refer to concepts or events; they have no physical referents e.g. SeasonOfYear attribute), Concrete Time (Refer to concepts or events; they have a sense e.g. Laps) and Learning Time (Refer to concept; the learner has learning time affected to him and it will be calculated each month).

The semantic description in ontologies can be a powerful tool in order to connect the interrelationship between abstract, concrete and learning time as LOs (Learning Objects); an example of that (has beginning, has duration, has time, has end etc.), this semantic description is more significant in daily learner life.

The Fig. 2 shows a brief taxonomy of concept and their semantics relationship:

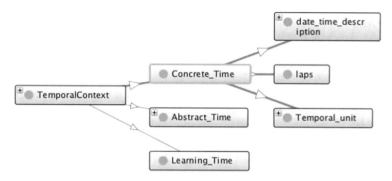

Fig. 2. Temporal context ontology

4.3   Learner Context Awareness

Learner play a major role as actor in an adaptive learning environment, each learner is characterized by different attributes related to his profile. We aim to propose a metadata that describe learner profile as following:

- *Basic information*: First name, Email, Tongue mother, etc.
- *Dynamic information*: Company, Institution, Job title, Learning style, Interest, Heath, etc.
- *User Type*: Learner, Author, Administrator, Etc.

The Fig. 3 shows a brief taxonomy of concepts:

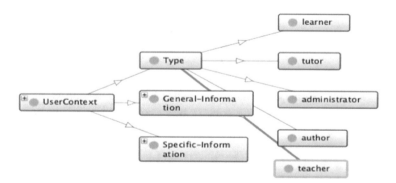

Fig. 3. User context ontology

4.4   Device Context Awareness

In order to adapt learning content to mobile devices, it's necessary to know the physical properties of this device.

Mobile device technology is characterized by Portability, Small Size, GPS, Wireless Communication, Battery Life, etc.

A remarkable study on device description ontology have been published such as WURF (Xml file), composite capability/preference (CC/PP) (RDF use), W3C 2007, Fipa device ontology (OWL ontology) UAU prof (user agent profile).

Under the name Device Context, we propose our ontology based on three categories:

- *GeneralDeviceInformation*: this category shows general information related to the device such as device name, device type, vendor.
- *HardwareDeviceInformation*: this category collects hardware information about device used by the learner such as CPU, Memory.
- *SoftwareDeviceInformation*: this category collects software information such as OS version, web browser, apps installed.

All information collected is used with user permission and it's used for adaptation purpose.

The Fig. 4 shows a brief taxonomy of concept and their semantics relationship:

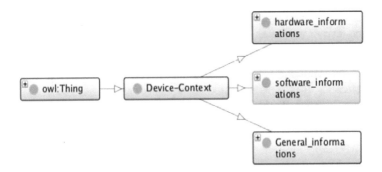

Fig. 4. Device context ontology

4.5   Vector Context Awareness

In order to model a learner context ontology, all model cited previously are grouped in a vector of context aware. A vector context aware contain all RDF objects.

Let's take C as learner context, a vector of context can be noted like:

$$V(c) = \{X(s), X(t), X(L), X(d)\}$$

$$X(S) : Spatial\,Context \quad X(T) : Temporal\,Context$$
$$X(L) : Learner\,Context \quad X(D) : Device\,Context$$

Preview Model: V(c) = ((Bus, Public, Noisy, Tangier, Morocco), (2017-14-7,10:00 pm, Monday, 1H), (Level_1, English,27, Center_of_interest), (Smartphone,4G, Android 7.0,30% Battery life, apps installed")).

The Fig. 5 shows a brief taxonomy of models:

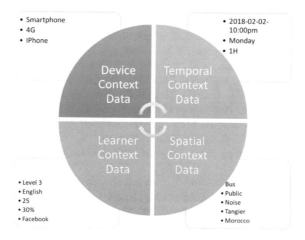

Fig. 5. Vector context ontology

5 Conclusion

In this paper, we introduce an ontological approach for service adaptation in context-aware mobile learning environment. therefore, we proposed, firstly, a context model which is generic and open to allow its extension to various changes based on learner needs. Then we perform a model execution in order to generate an OWL ontology that represents learner context aware.

The generated OWL ontology is then used to extract high level adaptations situations. also, we have proposed a semantic context-aware service, represented by extending OWL-S. This approach uses the power of ontologies and semantic web to represent and reason about context and perform service adaptation to learner needs according to their context.

Currently we are evaluating our ontological approach and we are completing the architecture of our adaptation Tool.

References

1. Aziz, A.K., Khoukhi, F.: A design requirements framework for mobile learning environment. Adv. Inf. Technol. Serv. Syst. 208–215 (2018)
2. Dey, A.K.: Understanding and using context. Pers. Ubiquitous Comput. **5**, 4–7 (2001)
3. Economides, A.A.: Context-aware mobile learning BT - the open knowledge society. a computer science and information systems manifesto. Presented at the Proceedings First World Summit on the Knowledge Society, WSKS 2008, Athens, Greece, 24–26 September 2008

4. Chen, H., Perich, F., Finin, T., Joshi, A.: SOUPA: standard ontology for ubiquitous and pervasive applications. In: Proceedings of MOBIQUITOUS 2004 - 1st Annual International Conference on Mobile Ubiquitous System Network Service, pp. 258–267 (2004)
5. Strang, T., Linnhoff-Popien, C.: A context modeling survey. In: Workshop on Advanced Context Modelling, Reasoning and Management at UbiComp 2004 - Sixth International Conference on Ubiquitous Computing Workshop, pp. 1–8 (2004)
6. Kapitsaki, G.M., Prezerakos, G.N., Tselikas, N.D., Venieris, I.S.: Context-aware service engineering: a survey. J. Syst. Softw. **82**, 1285–1297 (2009)
7. Wang, X.H., Zhang, D.Q., Gu, T., Pung, H.K.: Ontology based context modeling and reasoning using OWL. In: Proceedings of the Second IEEE Annual Conference on Pervasive Computing and Communications Work, pp. 18–22 (2004)
8. Gómez, S., Zervas, P., Sampson, D.G., Fabregat, R.: Context-aware adaptive and personalized mobile learning delivery supported by UoLmP. J. King Saud Univ. - Comput. Inf. Sci. **26**, 47–61 (2014)
9. Mantovaneli Pessoa, R., Zardo Calvi, C., Pereira Filho, J.G., Guareis de Farias, C.R., Neisse, R.: Semantic context reasoning using ontology based models BT - dependable and adaptable networks and services. Presented at the 13th Open European Summer School and IFIP TC6.6 Workshop, EUNICE 2007, Enschede, The Netherlands, 18–20 July 2007
10. Gao, Q., Dong, X.: A context-awareness based dynamic personalized hierarchical ontology modeling approach. Procedia Comput. Sci. **94**, 380–385 (2016)
11. Sagaya Priya, K.S., Kalpana, Y.: A review on context modelling techniques in context awarecomputing. Int. J. Eng. Technol. **8**, 429–433 (2016)
12. Krummenacher, R., Strang, T.: Ontology-Based Context Modeling. Ieice Trans. Inf. Syst. E90–D, 1262–1270 (2007)
13. Hayes, P. (ed): RDF Semantics: W3C Recommendation, 10 February 2004. http://www.w3.org/TR/2004/REC-rdf-mt-20040210/
14. Gu, T., Pung, H.K., Zhang, D.Q.: A service-oriented middleware for building context-aware services. J. Netw. Comput. Appl. **28**, 1–18 (2005)
15. Chen, H.L.: COBRA : an intelligent broker architecture for pervasive context-aware systems. Interfaces (Providence) **54**, 129 (2004)
16. Jeng, Y.-L., Wu, T.-T., Huang, Y., Tan, Q., Yang, S.J.H.: The add-on impact of mobile applications in learning strategies: a review study. Educ. Technol. Soc. **13**, 3–11 (2010)
17. Kumar, K., Saravanaguru, R.A.K.: Formalizing context aware requirement patterns using ontology. J. Teknol. **78**, 13–19 (2016)
18. International Standards Organization: ISO 8601:2004(E) Data elements and interchange formats - Information interchange - Representation of dates and times (2004)
19. Pan, F., Hobbs, J.R.: Time in owl-s. In: Proceedings of AAAI Spring Symposium Semantic Web Service, pp. 29–36 (2004)
20. Allen, J.F., Ferguson, G.: Actions and events in interval temporal logic. J. Log. Comput. (1994)

Structuring Areas of Interest in Learning Process as a Gaze Tracking Indicator

Asmaa Darouich$^{(\boxtimes)}$ and Faddoul Khoukhi$^{(\boxtimes)}$

Computer Laboratory of Mohammedia (LIM),
Faculty of Sciences and Technology Mohammedia (FSTM),
Hassan II University of Casablanca, 20650 Mohammedia, Morocco
asmaa.darouich@gmail.com, khoukhif@gmail.com

Abstract. The development of S-CAMO a decision making system for adaptive learning in real time, based on the study of the learner's profile is characterized by his/her gaze traces associated to each visited content. In this paper we propose a mathematical modeling approach that frames the construction of areas of interest defined by the spatial density of fixations based on the *dbscan* algorithm. However, the structuration of this model is related to the displayed course architecture, and is characterized by a scalable volume flow of gaze data.

Keywords: Gaze/Ocular data · Area of interest · Adaptive learning

1 Introduction

Evaluation of a learner is based on the analysis of interactions on accessing learning platform or applying activity theory on designing adaptive e-learning systems [1]. Eye tracking applied in learning systems is the process of tracking a learner eye/gaze during a learning process. Eye movements can be classified into three main categories [2]: fixations, during which the eye fixes a point, the scan path during which the gaze user follows an object and saccades which are a quick movements corresponding to the shift in interest an area to another one.

The main purpose of measuring eye/gaze movements in a decision making system for adaptive learning is to observe how learners visual attention is directed when they visit a web page in their learning process. In our approach, the proposed method to assess visual attention and areas of interest as a cognitive processes related to research information, take part of fixations that represent the retina stabilization on object. The development of such decision making system allows the personalization of learning resources to a corresponding learner profile and therefore make any adjustments. However, in our S-CAMO system (Fig. 1) we integrate the conception of a flexible learning content to reach a structured components' planning and scheduling. The content module deploys the implementation of the content by the tutor. As well as the provision of courses to learners. A course is a structure of a multiple combination of specific learning objects combined in different ways for semantically forming a pattern content.

© Springer Nature Switzerland AG 2019
F. Khoukhi et al. (Eds.): AIT2S 2018, LNNS 66, pp. 153–158, 2019.
https://doi.org/10.1007/978-3-030-11914-0_17

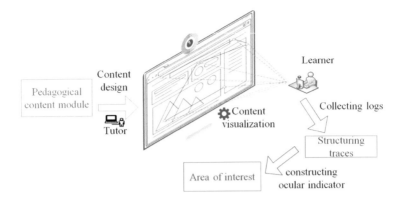

Fig. 1. S-CAMO system

Where a course web page is a pedagogical components Cpt_h structured by a set of resources R_p^h. An R_p^h could be a definition, an example, a theorem, a demonstration, a descriptive text, a study case, a problem solving, or a link back to a web page. Taking form as a text, an image, a sound, an audio, a video, an animation [3].

Based on the temporal fusion dynamics, each ocular data is merged with its browsing data accessed at time t_i that marks a pedagogical component Cpt_h [4]. However, an ocular data is a fixation $fix(t_j)$ as a gaze point on a pedagogical component Cpt_h.

$$fix\,(t_j) = (t_j, (x_j, y_j))$$

t_j: reflects the timestamp, ie the moment when the gaze point is captured on the screen.

x_j: is the abscissa of gaze point on the screen.

y_j: is the ordinate of gaze point on the screen.

When a fixation appears on an educational resource R_p^h, its coordinates are then limited by those of this resource (Fig. 2).

$$\left(X_{min}^p, Y_{min}^p\right) \le (x_j, y_j) \le \left(X_{max}^p, Y_{max}^p\right)$$

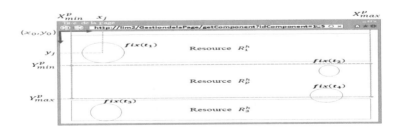

Fig. 2. Web page displaying resources to be learned at the system S-CAMO

2 Area of Interest Defined by the Spatial Density of Fixations

We propose the $DBSCAN_{ZI}$ algorithm (Fig. 6) which aims to build areas of interest carried by a learner on a pedagogical component.

The construction of areas of interest is based on the *dbscan* algorithm, which is a density-based data partitioning algorithm, in that it relies on the estimated cluster density to perform partitioning.

The dbscan algorithm uses two parameters: the *eps* distance and the minimum number of *MinPoints* points that must be in a *eps* radius for these points to be considered as a grouping.

The base is to manage the aberrant fixations by eliminating them from the partitioning process.

To find an area of interest zi_m, the *dbscan* algorithm starts by scanning the fixations vector $[fix(t_1) \cdots fix(t_m)]_h$. It searching for all accessible density points from the current fixation $fix(t_j)$. If this current fixation is a central point, then it will be added to the area of interest zi_m.

If $fix(t_j)$ is a noise point then no point is reachable from $fix(t_j)$ and the algorithm will visit the next fix $fix(t_{j+1})$ of the vector of fixations.

By using the global values *eps* and *MinPoints*, *dbscan* can merge two areas of interest in the case where these areas of interest of different density are close to each other. Two areas of interest of fixations with at least the smallest density will be separated from each other if the distance between them is wider than *eps*.

As a result, a recursive call to *dbscan* may be needed for clustering detected with the highest value of *MinPoints*. This is not necessarily a disadvantage because the recursive application of *dbscan* remains a basic algorithm and is only necessary under certain conditions.

Thus, for each added fixation, there is a growth zone that will make it possible to extend the zone of interest. Obviously, the larger the area, the more likely it is to expand. The concept of neighborhood is the key to this method, so we form the area of interest gradually.

Inputs for the Algorithm

- $eps = 100px$
- $MinPoints = 3$
- $R[fix(t_j)]$: represents the grouping associated with the fixation $fix(t_j)$ being processed.
- zi_q: The maximum set of connected points.
- Minkowski's distance: $D(fix(t_j), fix(t_r)) = \sqrt{(x_i - x_r)^2 + (y_i - y_r)^2}$.
- Diameter of an area of interest: $Diameter(zi_q) = \sqrt{\dfrac{\sum_{i=1}^{c} \sum_{j=1}^{c} \sqrt{(x_i - x_j)^2 + (y_i - y_j)^2}}{c}}$. c represents the number of fixations included in the area of interest concerned.
- Radius of an area of interest: $R(zi_q) = \dfrac{Diameter(zi_q)}{2}$.

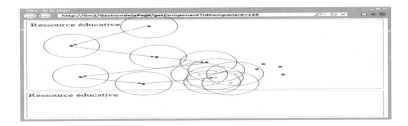

Fig. 3. Scan the current area of interest

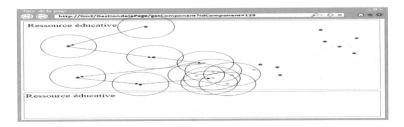

Fig. 4. Recovery of fixations in the neighbourhood of fixations of the current area of interest

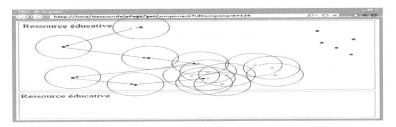

Fig. 5. Assigning fixations to the current area of interest

- Area occupied by an area of interest: $S(zi_q) = \pi * R(zi_q)^2$.
- Density of an area of interest: $Density_{zi_q} = \dfrac{R(zi_q)}{S(zi_q)}$.

The concept of construction of areas of interest from the fixations treated since the vector of fixations (Figs. 3, 4, and 5).

- The blue points are visited but with which the *eps* radius does not contain the number of points required to start a grouping, we move from one point to another according to their temporal scheduling.
- Green points are the points in a grouping from which, you re-assign more points in the *eps* range as part of expanding this grouping, or simply the starting point of a grouping.
- The red points are either noise point or not yet processed points.

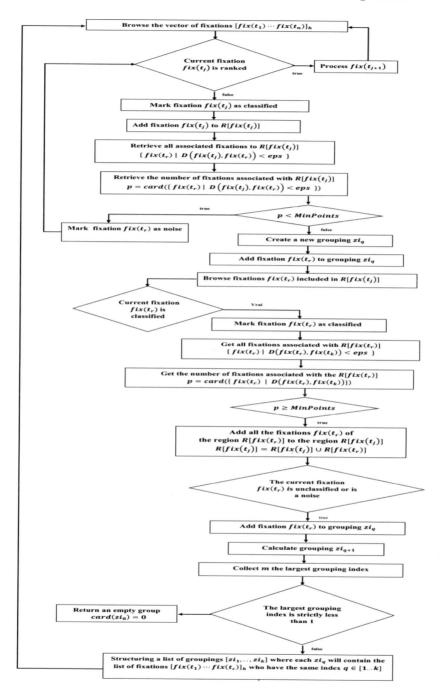

Fig. 6. The $DBSCAN_{ZI}$ algorithm for building areas of interest

3   Conclusion

The areas of interest in gaze tracking field represent the detection of areas that aim to reflect where the learner interest is on the visualised web page. The challenge is then to structure these areas of interest and to trace where they are projected at the level of a pedagogical component Cpt_h.

References

1. Peña-Ayala, A., Sossa, H., Méndez, I.: Activity theory as a framework for building adaptive e-learning systems: a case to provide empirical evidence. Comput. Hum. Behav. **30**, 131–145 (2014)
2. Freitas, S., Levene, M.: Wearable and mobile devices. In: Ghaoui, C. (ed.) Encyclopedia of Human Computer Interaction, pp. 706–712. Idea Group Reference (an imprint of Idea Group Inc.), Hershey, London (2005)
3. Darouich, A., Khoukhi, F., Douzi, K.: A dynamic learning content pattern for adaptive learning environment. In: 10th International Conference on Intelligent Systems: Theories and Applications (SITA), pp. 1–6. IEEE, Rabat (2015)
4. Darouich, A., Khoukhi, F., Douzi, K.: T-CAMO a traces model for adaptive learning environment. In: Mediterranean Conference on Information & Communication Technologies 2015, Saidia (2015)

Advances Natural Language Processing and Applications

Use SMO SVM, LDA for Poet Identification in Arabic Poetry

Alfalahi Ahmed[1]([⊠]), Ramdani Mohamed[1], and Bellafkih Mostafa[2]

[1] Département d'informatique, FSTM Université Hassan II Mohammedia,
Mohammedia, Morocco
flahi79@gmail.com, moha@fstm.ac.ma
[2] Institut National des Postes et Télécommunications, INPT-Rabat,
Rabat, Morocco
bellafki@inpt.ac.ma

Abstract. This work study a poet identification in Arabic poetry using classification methods, with features like poetry features, Sentence length, Characters, word length, first word in the poetry sentence and specific words are used as input data for text classification algorithms: Sequential Minimal Optimization (SMO), Support Vector Machine (SVM), and Linear discriminant analysis (LDA). The data set of experiment divided into two parts: a training dataset with known poets and test dataset with unknown poets. In our experiment, a set of 114 poets from entirely different eras are used. The researcher shows exciting results with a classification accuracy of 98.2456%.

Keywords: Arabic poetry · Poet identification · SMO · SVM · LDA

1 Introduction

Arabic poetry is the art earliest shape of Arabic literature. Traditionally, this poetry is classified into two sets: measured or rhymed, and poem prose. The measured or rhymed poetry is considerably preceding the latter since they appear historically eeliest. The rhymed poem is classified by fifteen different meters. Such meters of the measured poetry is also, known in Arabic as "seas" (بحور/bohour). As mentioned before, the syllables are measuring unit of seas "tafilah" (تفعيلة). Each sea contains a certain number of "taf'ilah" which the poet has to observe in every verse (بيت/bayt) of the poem. The procedure of counting some taf'ilah in a poem is stringent since adding or removing a consonant or a vowel letter "حركة" can shift the bayt from one meter to another. Another feature of measured poems is that every bayt (the second part of the verse) end with the same letter (rhyme) throughout the poem (Wiki 2015).

The task of analyzing the text content to identify its original poet among a set of candidate poets is called Author identification. The concept at the back of the author identification is as follows: given a collection of poetry texts as training dataset of the known poet. The unchecked text is decided via matching the nameless textual content to at least one poet of the candidate set agreed via matching the unknown textual

---

F. Khoukhi et al. (Eds.): AIT2S 2018, LNNS 66, pp. 161–169, 2019.
https://doi.org/10.1007/978-3-030-11914-0_18

content to at least one poet of the candidate set. In the context of Classical Arabic poetry, the current task can be re-formalized as follows:

Given poetry with an anonymous author, find to whom this poem is belonging known set of features of each candidate authors. Author identification research in Arabic poetry is considered new and is not tackled as much as in other languages (Shaker and Corne 2010). Before this research, no published works and researchers about author identification in Arabic poetry except our paper (Ahmed et al. 2016; Al-falahi et al. 2017). The most founding research deal with Arabic poetry as a classification task. Al Hichri and Al Doori, in (Alhichri and Aldoori 2008) used the distance-based method to classify Arabic poetry depending on the rhythmic structure of short and long syllables. The Naïve Bayesian classifier was used by Iqbal Abdul Baki (Mohammad 2009) to organize the poetry into classification sets known in Arabic as " meters" (buḥūr).

Alnagdawi and Rashideh (Alnagdawi et al. 2013) proposed a context-free grammar-based tool for finding the poem meter name. The system introduced was worked only with trimmed Arabic poetry (words with diacritics). On the alternative hand, there are little works address author identification of Arabic (Abbasi and Chen 2005; Frantzeskou et al. 2007; Shaker and Corne 2010; Altheneyan and Menai 2014; Baraka et al. 2014). Among them, Altheneyan and Menai's work (Altheneyan and Menai 2014) is attractive. In their work, four different models of Naïve Bayes classifiers: multinomial Naïve Bayes, simple Naïve Bayes, multi-variant Poisson Naïve Bayes, and multi-variant Bernoulli Naïve Bayes were used in his work. The experiment becomes especially depending on feature frequency that is extracted from a massive corpus of 4 different datasets. The average results confirmed that the multi-variant Bernoulli Naïve Bayes version gives the best results among all used versions because it changed into capable of locating an author of textual content with a mean accuracy of 97.43%.

Markov chains use in some works using but are not new in this direction (Kukushkina et al. 2001; Khmelev and Tweedie 2002), however, using SMO, SVM or LDA together are new in our attempt to apply them together in Classical Arabic Poetry context. Current paper proposes to use, SMO, SVM, and LDA to solve author identification in Arabic poetry.

Thus, the organized of paper like the follows: Sect. 2 showed a general overview of the characteristics of Arabic poetry. Section 3 introduces the Arabic Poetry Corpus. Section 4 presents the Author identification methodology. Section 5 discusses the results of the test work. Finally, a conclusion and future work are present in Sect. 6.

2   Arabic Poetry Characteristics

Classical Arab poetry that includes some of the characteristics that distinguish it from the rest of literary, it's called Rhyme, Meter, and shape.

2.1 Rhyme (Alqafiyah)

The process of finding the rhyme of Arabic poetry is easy. In traditional Arabic poetry, poetry follows a rigorous but straightforward rhyme model (Ahmed et al. 2015). Since the end letter of each verse in traditional Arabic poetry must be the same letter. The rhyme is the give up a letter of the second one part of any verse. In the case of vowel letter, then the second remaining letter of every verse ought to be the same as nicely. The vowel sounds in Arabic are a "ا" alif, i "ي" yaa, and o "و" wow. Each vowel sound has two versions: a long and a quick model. Short vowels are written as diacritical marks below or above the letter that precedes them even as long vowels are written as whole letters (Almuhareb et al. 2013).

2.2 Meter (Alwazn)

Traditional Arabic Poetry has a restricted structure that is in particular based entirely on the length of syllables. This shape formulates, as stated before, the meter. Traditionally, there are seventeen meters defined using the grammarian al-Khalili in the 8th century. Each meter is produced from fundamental gadgets referred to as 'peg' (watid) and 'cord' (sabab). Each group consists of either short or long syllable (Scott 2009).

3 Corpus of Arabic Poetry

The corpus of Arabic poetry may be a store having an assortment of poetry related to a specific poet. The poetry of totally (114) different poets is collected from some websites and poetry books (Table 1).

The poetry corpus in this work includes 114 poets with 12978 Qasidah, the full words is 12311402 divisions to 174826 words for training dataset, and 60576 words for the testing dataset is shown in Table 2.

Table 1. The corpus of Arabic poetry details

	Name poet	N Qasidah	N verse	N meter	N words training	N word test
1	Shoqi	51	366	8	4410	468
2	IbenHani	134	1760	11	17914	797
3	Motanabi	365	4050	15	57117	870
114	Abutammâm	313	7256	13	17168	682

Table 2. The corpus of Arabic poetry

	N. poets	N. Qasidah	N. words
Training dataset	114	12978	1174826
Testing dataset	114		60576
Total	228	12978	12311402

4   Poet Identification Methodology

To build our poet identification model (Fig. 1), the researchers go through some steps: text preprocessing, extraction of features and features selection for Poetry Author fingerprint detection. In this work, the researchers present the Author identification task as a classification process. The methodology the researchers applied starts from a classification of a preprocessed data. The dataset is divided into test datasets and training dataset. In the first phase, preprocessing for text and determine prescient features are extracted from the dataset, then the training and test datasets are made, on the premise of these features. The second step, building the model to process the training data, and test data, then it is tested on text unknown. The test and training cases are numerical features vectors that represent term frequency of each chose features, taking after by the poet's name. The researchers perform administered class, the situation in which named training data are utilized to train a system learner, because it permits the evaluation of classification, and for this reason is the first-rate approach for examining the adaptability of the Text Classification technique (Luyckx 2010).

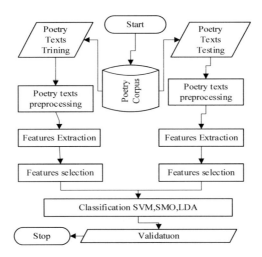

Fig. 1. Poet identification model

4.1   Text Preprocessing

The researchers collected the poets' texts of the study sample from poetry encyclo-pedias and websites, for some famous poets. The poets have been choosing randomly from specific eras. The bulk of the poetry texts are utilized in training data. However, the ultimate used as testing data. The researchers have introduced a range of 114 unknown poetry texts varying the number of verses (abyat/ابيات) of the test. The collected texts were classical poetry which relies on weight and rhyme and not pure. They consisted of some alphanumeric and punctuation that is typically laid out in such poetry type. Thus, maximum of the poetic texts are difficulty to initialization system of strip punctuation, strip numbers, and strip alphanumeric. In the system of normalization, some Arabic letters have distinct paperwork inclusive of (إ, أ, آ) to (ا) given that they do now not deliver any indication of discriminatory classical poetry texts, however it can play an essential function in a few established texts inclusive of net pages (Howedi and Mohd 2014).

4.2 Extracting Features

This step of extracting features for author identification is a critical phase to discover distinct features. For each poet, the researchers impose that the poet has specific style features. The researchers can recognize four principle sorts of features that convey possible indicators for authorship: syntactic, character, semantic, and lexical features. In this paper, the researchers give an account of trials utilizing lexical and character features, since they are more robust than semantic features, considering the cutting edge in the semantic validation; and most regularly connected behind syntactic features. The features the researchers utilized are records as a part of Table 3. Sentence length, rhyme, meter, word length, first-word, and Characters, in a poetry sentence, length have been exhibited that they find themselves able to handle restricted data dependably (Luyckx 2010).

4.3 Feature Selection

When all features are extracted in the previous step, the feature selection is applied to limit the dimension of potentially relevant features. Feature choice is a vital a part of each author identification study, which aims to identify the most relevant ones for the task. The frequency of a feature is the most used criterion for selecting features for author identification. The simplest way is to restrict the set to the *n* most frequent terms in the dataset (Howedi and Mohd 2014). The point of features choice routines is to diminish the dimensionality of the dataset by uprooting unessential features for the grouping undertaking. A few sorts of features, for example, character and lexical features can impressively expand the dimensionality of the features set (Stamatatos 2009).

In such case, features determination techniques can utilize to reduce such dimensionality of the frequency. When all in achieved, the greater continuous features are, the more elaborate variety, it catches. In this work, the features are pick by using two well-known characteristic choice methods: chi-squared (χ^2) as (Zhao and Zobel 2005) and information gain (IG) methods as (Howedi and Mohd 2014).

4.4 Experiment

In our experiment and after extracting feature values using "Rapidmenar tool" the researchers group them into seven sets according to following authorship features: F1 set of character features, an F2 set of word length, F3 set of sentence length, F4 set of first-word in a sentence, F5 set of the Meter and F6 set of rhyme, F7 set of Specific words. Studies have confirmed that lexical and syntactic features are the most vital classifications and therefore structure the status quo of auxiliary and substance specific features (Stamatatos 2009). The researchers connected this idea to test the importance of classification features of Arabic poetry. For the experiment, the researchers created 114 randomly selected samples of 114 authors, which the researchers used in all experiments.

The researchers evaluated each sample using most poetry text per author on training data and conducted a 114-unknown text to the detected author with SMO, SVM, LDA classifiers, which the researchers used a part of all investigations. The accuracy of a category model is printed in the time period of traditional accuracy (a whole variety of adequately known text over the 114 complete texts).

5 Results and Discussion

In our many experiments of author identification on old Arabic poetry, a set of texts that were written by 114 Arabic poets are introduced. Many features: words length, sentences length, characters, the first word in sentences, meter, rhyme, and specific words are tested (Ahmed et al. 2015). The researchers observed from Table 3 after applying all algorithms that maximum accuracy value is 98.2456% of correct attribution with apply LDA. The accuracy is used to quantity demonstrate of appropriately characterized examples over the mixture number of test cases by means of figuring the normal of accuracy.

Table 3 demonstrates the best features score that acquired in utilizing SMO, SVM, LDA. The maximum value is attained by applying LDA on specific word features (accuracy = 98.2456%), this means that the specific words in Arabic poems used in different ways by the poets and can distinguish among poets stylistic. The same value (98.2456%) the researchers obtained when implementing LDA on a combination of F1 + F2 + F3 + F4 + F5 + F6 + F7 features in Table 4. We found the best result for all the features that have used in our experiment. However, the researchers obtained the least result 71.929% by applying SVM on F4 = first word features (Table 3) and nearest value 72.807% by applying SVM on a combination of F1 + F2 + F3 + F4 + F5 features.

The small rate 71.929% cannot use to certify that the correct poet wrote a text through the first-word length because this feature does not repeat with all poets. The next small rate 74,561% from SMO was intended for a data handle with different cases with Sentences length features; this means that the Sentences length poetry is not a clear sign of perceiving the authors of the texts because the similarity with the meter determines Sentences length by most poets. Nevertheless, when the researchers used meter feature and rhyme feature in our experiment the researchers obtained maximum rate 87,719% for rhyme and 78.947% for the meter in Table 3 this means that rhyme and meter features given a good result when the researchers applied LDA and same value (78.947%) obtained by SMO, SVM on rhyme feature. Likewise, a score of great attribution of 97.368% through using one of the compact two features: The F1 + F2 via F1when the researchers applied SMO on Table 4.

Likewise, the researchers obtained the same result of the experiment on features F2 + F1 word length and characters together, it is a good result compared with the result of F4 + F3 + F2 + F1, which added word length F4 and sentence length F3 to F2 + F1 where the value is 74.561%, 74.561% and 77.192% used SMO, SVM, and LAD.

This decline because of first word length and the sentence length in old Arabic poetry obligate by some "taf'ilah" and meters (wazn).

The researchers also observed that the sentence length (74.561%, 74.561%, 92.105%) and first word length (73.684%, 71.929%, 77.1929%) when tested separately by used SMO, SVM, and LDA while when added to the other features was varying according to integrating with different features.

Table 3. The accuracy of proper attribution obtained the different features

Features	Total correct			Accuracy percent		
	SMO	SVM	LDA	SMO%	SVM%	LDA%
Character	111	105	108	97.3684	92.1052	94.7368
Word length	101	106	112	88.5964	92.9824	98.2456
Sentences length	85	85	105	74.5614	74.5614	92.1052
First word length	84	82	88	73.6842	71.9298	77.1929
Meter	86	87	90	75.4385	76.3157	78.9473
Rhyme	89	90	100	78.0701	78.9473	87.7192
Specific words	112	111	112	98.2456	97.3684	98.2456
Average	95.4285	95.143	102.1426	83.7092	83.4586	89.5989

Table 4. The accuracy attribution obtained the different features together

Features	Accuracy percent					
	SMO	SMO%	SVM	SVM%	LDA	LDA%
F1, F2	111	97.3684	103	90.3508	112	98.2456
F1, F2, F3	101	88.5964	106	92.9824	105	92.1052
F1, F2, F3, F4	85	74.5614	85	74.5614	88	77.1929
F1, F2, F3, F4, F5	86	75.4385	83	72.8070	112	98.2456
F1, F2, F3, F4, F5, F6	104	91.228	108	94.7368	106	92.9824
F1, F2, F3, F4, F5, F6, F7	111	97.3684	112	98.2456	112	98.2456
Average	99.6666	87.4269	99.5	87.2807	105.833	92.8362

6 Conclusion

In our work, an Author identification task has experimented on old Arabic poetry set of texts that written by 114 Arabic poets. Each author was present by many different texts. The algorithms classifiers are implemented on many texts validation. The experiments, which have done separately for each feature on the old Arabic Poetry dataset using SMO, SVM, LDA classifier shows the following remarkable points.

- The F7 = Specific words, F1 = Character, and F2 = word length features are better than all features in Table 3, regarding the maximum accuracy = 98.2456%, 98.2456%, 97.364% by LDA, SMO, SVM on specific words and maximum accuracy = 98.24% by LDA on word length and accuracy = 97.368% by SMO of total features.

- The meter without other features does not give a clear indication contributed to accurate poet identify but while use in other features is giving a good result.
- Meter and Rhyme features are fewer results and cannot be used to certify that correct Poet wrote a text if it used separately.
- The best performance the researchers got after it has been used all features together regarding the accuracy = 98.2456%, 98.2456%, 97.368% when the researchers use LDA, SVM, SMO on these features.
- Rhyme with other features gives a clear indication contributed to accurate author identify but while used alone not giving a good result.
- The best Average the researchers obtained from applied LDA algorithm on all features is 92.836% better than other algorithms.

7 Future Work

The researchers propose to use other poetry features and used some features like weight, synonyms and rare word to overcome those obstacles.

In addition, the researchers propose a plan to extend the investigations into bigger datasets of poets. The researchers recommend to extend experiments using other algorithms and compared the results with these new results.

Acknowledgments. This paper is supported by IBB University in Yemen. The researchers tend to convey thanks to our colleagues from FSTM and INPT who provided insight and skill that greatly motor-assisted the paper.

References

Abbasi, A., Chen, H.: Applying authorship analysis to Arabic web content. In: Kantor, P., et al. (eds.) Intelligence and Security Informatics, pp. 183–197. Springer, Heidelberg (2005)

Abbasi, A., Chen, H.: Applying authorship analysis to extremist-group web forum messages, September/October 2005

Ahmed, A., Mohamed, R., Mostafa, B.: Authorship attribution in Arabic poetry using NB, SVM, SMO. In: 2016 11th International Conference on Intelligent Systems: Theories and Applications (SITA), pp. 1–5. IEEE (2016)

Ahmed, A.F., Mohamed, R., Mostafa, B., Mohammed, A.S.: Authorship attribution in Arabic poetry. In: 2015 10th International Conference on Intelligent Systems: Theories and Applications (SITA), pp. 1–6 (2015)

Al-falahi, A., Ramdani, M., Bellafkih, M.: Machine learning for authorship attribution in Arabic poetry. Int. J. Futur. Comput. Commun. **6**, 42–46 (2017)

Alhichri, A.M.A, Aldoori, M.A.: Expert System for Classical Arabic Poetry (ESCAP). In: Proceedings of the International Conference on APL, Toronto, Ontario, Canada (2008)

Almuhareb, A., Alkharashi, I., AL Saud, L., Altuwaijri, H.: Recognition of classical Arabic poems. In: Proceedings of the Workshop on Computational Linguistics for Literature, pp. 9–16 (2013)

Alnagdawi, M.A., Rashideh, H., Fahed, A.: Finding Arabic poem meter using context free grammar. **3**, 52–59 (2013)

Altheneyan, A.S., Menai, M.E.B.: Naïve Bayes classifiers for authorship attribution of Arabic texts. J. King Saud Univ. Comput. Inf. Sci. **26**, 473–484 (2014). https://doi.org/10.1016/j.jksuci.2014.06.006

Baraka, R., Salem, S., Abu, M., Nayef, N., Shaban, W.A.: Arabic text author identification using support vector machines. J. Adv. Comput. Sci. Technol. Res. **4**, 1–11 (2014)

Khmelev, D.V., Tweedie, F.J.: Using Markov chain for identification of writers. **16**(4), 12 (2002)

Frantzeskou, G., Stamatatos, E., Gritzalis, S., Chaski, C.E., Howald, B.S.: Identifying authorship by byte-level n-grams: the Source Code Author Profile (SCAP) method. Int. J. Digit. Evid. **6**, 1–18 (2007)

Howedi, F., Mohd, M.: Text classification for authorship attribution using Naive Bayes classifier with limited training data. Comput. Eng. Intell. Syst. **5**, 48–56 (2014)

Kukushkina, O.V., Polikarpov, A.A., Khmelev, D.V.: Using literal and grammatical statistics for authorship attribution. Probl. Inf. Transm. **37**, 172–184 (2001). https://doi.org/10.1023/A:1010478226705

Luyckx, K.: Scalability issues in authorship attribution (2010)

Mohammad, I.A.: Naive Bayes for classical Arabic poetry. **12**, 217–225 (2009)

Scott, H.: Pegs, cords, and ghuls: meter of classical Arabic poetry (2009)

Shaker, K., Corne, D.: Authorship attribution in Arabic using a hybrid of evolutionary search and linear discriminant analysis. In: 2010 UK Workshop on Computational Intelligence (UKCI) (2010)

Stamatatos, E.: A survey of modern authorship attribution methods. J. Am. Soc. Inf. Sci. Technol. **60**, 538–556 (2009). https://doi.org/10.1002/asi.21001

Wiki: Arabic poetry-Wikipedia, the free encyclopedia (2015). http://en.wikipedia.org/wiki/Arabic_poetry. Accessed 15 Apr 2015

Zhao, Y., Zobel, J.: Effective and scalable authorship attribution using function words. Inf. Retr. Technol. **3689**, 174–189 (2005). https://doi.org/10.1007/11562382_14

Automatic Extraction of SBVR Based Business Vocabulary from Natural Language Business Rules

Abdellatif Haj$^{(\boxtimes)}$, Youssef Balouki, and Taoufiq Gadi

Department of Mathematics and Computer Science, LAVETE Laboratory,
FSTS, Hassan 1st University, Settat, Morocco
haj.abdel@gmail.com, balouki.youssef@gmail.com,
gtaoufiq@yahoo.fr

Abstract. In the early phases of software development, both of business analysts and IT architects collaborate to define the business needs in a consistent and unambiguous format before exploiting them to produce a software solution to the problem have been defined. Given the divergence of the interest areas of each intervenor, the natural language remains the most adequate format to define the business needs in order to avoid misunderstanding. This informal support suffers from ambiguity leading to inconsistencies, which will affect the reliability of the final solution. Accordingly, the Object Management Group (OMG) has proposed the "Semantic Business Vocabulary and Rules" (SBVR) standard which offers the opportunity to gather business rules in a natural language format having a formal logic aspect, letting the possibility to be understood by not only the different stakeholders but also directly processed by the machine. Since the SBVR standard is born to represent business rules by combining business vocabulary, it would be wise to give a great attention to the latter. In this paper we present an approach to extract business vocabulary according to SBVR Structured English as one of possibly notation that can map to the SBVR Meta-Model, with a view to provide a relevant resource for the next software deployment steps.

Keywords: Business rules · Business vocabulary ·
Semantic of Business Vocabulary and Rule · SBVR ·
Natural Language Processing · SBVR structured English

1 Introduction

The constant evolution of technology is the major cause of software development crisis, namely, complexity and ability to change. Software development relaying on source code had become a bad choice for developers, as it often requires theme to rewrite the entire code to ever-changing of business rules which is costly to businesses. Accordingly, The Object Management Group (OMG) [4] has adopted on 2001 the Model Driven Architecture (MDA) as a solution to this problem based on separation of concerns. MDA allows the separation of the business aspect and the technical aspect of a system using models at each step of the development process, leading to automatic

© Springer Nature Switzerland AG 2019
F. Khoukhi et al. (Eds.): AIT2S 2018, LNNS 66, pp. 170–182, 2019.
https://doi.org/10.1007/978-3-030-11914-0_19

generation of source code as a final step. Clearly then, models – the backbone of the MDA process – are considered not just as a communication support but rather a productive element, which can accelerate not only the production of software solutions but also its maintenance besides its reusability.

Habitually, the software modeling begins with an elicitation phase to identify business requirements before analyzing and producing the source code. MDA follows this trend and defines three levels of models, representing the abstraction levels of the system: the CIM, PIM and the PSM model:

1. **CIM** (Computational Independent Model): To describe the functional requirements of the system (the services it offers and the entities with which it interacts).
2. **PIM** (Platform Independent Model): To describe the structural (static) and behavioral (dynamic) aspect of the system.
3. **PSM** (Platform Specific Model): To describe a particular platform for generating code source.

Clearly then, the CIM & PIM models describe the system independently of its implementation, while the generation of the source code is left as a final step in the PSM model. Since each model is based on his previous model; the reliability of each model will be related to the level of reliability of his previous model, which supports the fact that the inception step is considered as the most important phase of all the software development cycle.

Traditionally, analysts and IT architects use natural language in the early phases of developing enterprise-scale solutions. This informal support suffers from ambiguity which is extremely complex and difficult to automate as well. To deal with this problem, business IT tends to use formal language such as Alloy [5] or even a semi-formal one such as UML [6]. However, in the most of the cases, business architects are not familiar with (semi) formal languages, which can complicate the communication between different stakeholders coupled with erroneous comprehension that can affect the rest of the development process.

As a solution to this problem, Object Management Group (OMG) [4] proposed in 2008 "Semantic Business Vocabulary and Rules" (SBVR) standard [2], to capture business rules (BR) in a natural language (NL) form, which can be processed automatically due to its formal logic aspect. Such a solution can help simplifying communication and bridging the gap between business experts and engineers. Since its release, SBVR has become widely used to overcome natural language problems, such as ambiguity and machine processing, therefore, increase the accuracy of approaches dealing with informal language.

SBVR – as it is clearly understood from the name – standardizes two important areas of semantics: (1) Business Vocabulary (BV) dealing with the meaning of all kinds of concepts used by a given organization; (2) Business Rules (BR) dealing with the meaning of the business rules with regard to the business vocabulary, which realizes the Business Rules Mantra: "Rules build on facts, and facts build on terms" [1]. That is to say, the business vocabulary on which all rules are based has an essential role in supporting business rules.

In this paper, we propose an approach to extract the business vocabulary according to SBVR standard, from rules written in a natural language using Natural Language Processing (NLP), in order to create a strong base for Business rules, following the steps outlined in Fig. 1:

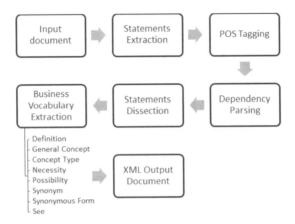

Fig. 1. Overview of our approach.

1. Extract natural language business statement.
2. Extract grammatical information of each statement such as POS tagging and dependency relation.
3. Mapping extracted element to SBVR concept.
4. Extract information about each concept.
5. Save extracted information in a XML file.

SBVR Structured English (SBVR-SE) notation [2] (Annex A) will be used in this paper – due to its popularity – as one of possibly notation mapping to the SBVR Meta-Model.

This paper is organized as follows: The next section gives a brief overview of SBVR Structured English notation. The detail of our approach is given in the third section. A case study illustrating the feasibility of our approach is given in the fourth section. Our conclusions and future works are drawn successively in the final two sections.

2   Related Works

Various approaches have been proposed to extract Business vocabulary and Rules from natural language specifications. Bajwa et al. [8] propose an approach for translating NL specification to SBVR using NLP, which is extended to generate OCL [9] and then to generate Alloy [10]. However, Business Vocabulary needs a manual effort to be created in form of UML model as a context of the input English. Umber et al. [11–13], Bajwa et al. [14], Ramzan [15, 16] also propose an approach to automatically translate the

English Requirements Specifications to SBVR, then again, they have been limited to extract basic SBVR elements, e.g. noun concept, individual concept, object type, verb concepts, etc. without any extra information. Roychoudhury et al. [17] present a (semi-) automated transformation of the legal NL (English) text to SBVR Model via authoring of Structured English (SE) rules. The approach uses a vocabulary editor for capturing concepts, terms, definitions, synonyms etc. Hypsky et al. [18] propose to define and formalize the business rules using an already specified business vocabulary and semantics. The approach uses a Business dictionary which must contain all the terms that are used within an organization. Afreen et al. [19], Thakore et al. [20], Mohanan et al. [21, 22], propose to generate UML model from Natural language via SBVR. Njonko [23] propose an approach to transform natural Language Business Rules to executable model (such as SQL). Some of previous works use UML model as a context of the input English to provide business Vocabulary, which not only involves a prior knowledge of UML, but also affects the automation of the process. Other works have been limited to extract basic SBVR elements but failed to generate extract information such as Definition, Synonym, hypernym or other attributes defined in SBVR Structured English notation, which will enrich the business vocabulary. Our contribution is to extract Business vocabulary automatically according to SBVR standard, from rules written in a natural language using Natural Language Processing (NLP), in order to create a strong base for Business rules.

3   SBVR Structured English

As mentioned earlier, we have adopted SBVR Structured English (SBVR-SE) notation [2] for mapping to the SBVR Meta Model, which proposes a description for both business vocabulary and business rules. For brevity, we will present only the part describing the vocabulary and fonts used by SBVR-SE, which are the subject of this paper.

3.1   SBVR-SE Fonts

SBVR-SE Statements are completely expressed using four fonts:

- Term: used for noun concept (usually in lower case letters) (e.g. customer).
- Name: used for individual noun concept (usually with capitalized first letter) (e.g. California).
- Verb: used for verb concepts (e.g. customer has name).
- Keyword: linguistic symbols (e.g. the, each, it is obligatory that).

An example of an SBVR statement can be seen in Fig. 2.

Fig. 2. Overview of SBVR elements written in SBVR-SE notation.

3.2   Vocabulary Description

SBVR-SE proposes a business document in the form of glossary entries, which is begin by giving an introduction to a vocabulary description according to the following:

<Vocabulary Name>

Description	: scope and purpose of the vocabulary
Source	: based on a glossary or other document
Speech Community	: responsible for the vocabulary
Language	: English by default
Included Vocabulary	: indicate another incorporated vocabulary
Note	: explanatory notes

Similarly, each single concept is presented according to the following:

<Primary representation>	: term, a name, or a verb concept wording
Definition	: expression defining the primary representation
General Concept	: a concept that generalizes the entry concept
Concept Type	: specify a type of the entry concept
Necessity	: to state that something is necessarily true
Possibility	: to state that something is a possibility
Synonym	: designation for the same concept
Synonymous Form	: synonym for verb concept wording
See	: preferred representation
Source	: source vocabulary for a concept
Dictionary Basis	: a definition from a common dictionary
Example	: labels examples
Note	: to label explanatory notes
Subject Field	: a qualification
Namespace URI	: indicate a URI of a vocabulary

More details on this topic can be found in [2] (Annex A).

The last five elements in the primary representation of concepts (source, Dictionary Basis, Example, Note, Subject Field, Namespace URI), as well as the global information about business vocabulary, are difficult if not impossible to extract directly from business rules due to the need of a human determination. Furthermore, those elements will weigh down the process vocabulary extraction without any contribution in the further automation of the development process. Hence, there extraction will be out of the scope of this work. Accordingly, we will limit ourselves to extract only pertinent information for each noun or verb concept, namely:

<Primary representation>
Definition
General Concept
Concept Type
Necessity
Possibility
Synonym
Synonymous Form
See

4 Our Approach

In this paper we highlight the automatic extraction of SBVR based business vocabulary from business rules written in English natural language. The extracted vocabulary will be saved according to SBVR-SE notation following the steps:

4.1 Statement Extraction

In the initial stage of the process, a document containing natural language business statements will be taken as an input file, in which, statements are separated by full stop, before saving each single statement separately.

4.2 POS Tagging and Dependency Parsing

Using "StanfordNLP" tool [3], each single statement will be POS tagged separately (Part-Of-Speech tagging) to extract grammatical information for each single word, such as Proper Nouns (NNP), Noun phases (NN) and verbs (VBZ) before generating the dependency grammar relation between words. An example can be seen in Fig. 3.

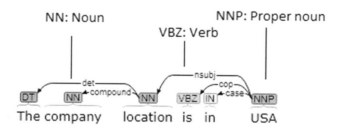

Fig. 3. Overview of POS tagging & dependency parsing of a statement.

4.3   Statements Dissection

Each statement will be analyzed and dissected in order to extract clauses composed each statement beside embedded information according to the following:

– Extraction of clauses:

A clause is composed of a set of words containing a verb with his subject, and can be either "independent" if it's a sentence standing on its own, or, "dependent" when it needs another independent clause to complete the meaning of the sentence.

Accordingly, all 'Verb Concept' will be extracted coupled with their 'Subject' and 'Predicate' (without determinant) according to the following pattern:

<Subject> <Verb> <Predicate> (e.g. customer has name).

If the sentence is in passive voice, then the name that follows 'by' keyword will be the 'Subject', where the 'Subject' of the passive voice statement will be the 'Direct Object'.

– Extraction of embedded information about 'Noun Concepts':

Embedded information about Noun Concepts will be extracted as illustrated in Table 1.

Table 1. Embedded Information about Noun Concepts

N°	Pattern	Description	Example
1)	Noun + Noun	'Noun Concepts' in compound relation.	NN ←compound→ NN Car renter
2)	Noun + Gerund + Direct Object	'Noun Concepts' having a 'Gerund' (VBG) in the form of an 'Adjective Clause' (acl) which in turn has a 'Direct Object' (dobj).	NN ←acl→ VBG ←dobj→ NN Car having renter
3)	Noun + ' of ' + Noun	'Noun Concepts' having noun modifier (nmod) joined with 'of' keyword.	—nmod— NN IN ←case→ NN Renter of car
4)	Noun + ' 's ' + Noun	'Noun Concepts' having possessive ending and a noun modifier.	—nmod:poss— NN ←case→ POS NN Car 's renter
5)	Adjective + Noun	'Noun Concepts' having adjectival modifier (amod).	JJ ←amod→ NN moral person

States from 1 to 4 will be added as a *characteristic verb concept* according to the following pattern: <Noun> <has> <Noun> (e.g. "car has renter").

For the last state (N° 5), it will be added as *categorization* of a noun concept according to the following pattern: <Noun> <is a category of> <Noun> (e.g. "moral is a category of person").

4.4   Extraction of SBVR-SE Based Business Vocabulary

The process continues by extracting a set of information about each concept following the rules outlined below:

'Definition' Caption
To give an automatic definition to our concepts, one can use any dictionary or thesaurus API such as Merriam-Webster [24], but to stay using our business vocabulary, we will rely on our business rules to extract definitions.

– For General Concept:

We will look for definitional statements providing specific characteristics or properties to our General Concept, so we look for statements having our General Concept in the subject part according to the following pattern:

<Our General Concept> is/are <Predicate>
<Our General Concept> has/have <predicate>

At this level, we are looking for providing a global definition for all General Concept without exception, that is why, we will ignore statements defining our General Concept coupled with an adjectival clause (using gerund), or an adverbial clause (using 'if', 'when'), or designation (using 'that', 'who', 'which').

– For Individual Concept:

Individual Concept such as the name of country, person or company, generally don't have to be defined, so we limit ourselves to put their general concept as a definition using StanfordNLP [3].

– For Verb Concept:

To give a definition to statements having an action verb (dynamic verb), we will give more details about it by looking for other statements having condition according to which our statement will be true. To this end, we will look for statements with verb having an 'adverbial clause' (that use 'if', 'when', …).

In the other hand, definitional statement having verb *'to be'* or *'to have'* describing a characteristic or specialization/generalization, will be defined according to Table 2:

Table 2. Pattern to describe non action verb.

Pattern	Verb concept (fact type)
<Subject> <to be> <Adjective>	<Adjective> 'is a characteristic of' <Subject>
<Subject> <to have> <Noun>	<Noun> 'is a characteristic of' <Subject>
<Subject> <to be> <Noun>	<Subject> specializes <Noun> <Noun> generalizes <Subject>

'General Concept' Caption

We will parse all extracted concept and compare each one with our General Concept using WordNet tool [7] to extract their hypernym (more general word) if exist, and put it in the 'General Concept' caption.

'Concept Type' Caption

The concept type caption can be extracted according to Table 3:

Table 3. Pattern to extract for concept type caption.

Pattern	Concept type
<Pronoun>	Individual concept
<Noun> <Action Verb><Noun>	Verb concept
<Noun> <to be> <Adjective>	Characteristic
<Noun> <to have> <Noun>	Characteristic
<Noun> <to be> <Noun>	Generalization
<Noun> <contain | include | incorporate> <Noun>	Partitive verb concept

'Synonym', 'Synonymous Form' & 'See' Caption

– Synonym:

We will parse all extracted *General Concept* and compare each one with our General Concept using WordNet tool [7] to detect its synonyms if exist, and put them in the 'Synonym' caption.

– Synonymous Form:

On the other hand, 'Synonymous Form' caption is designed for Verb Concept, so, for this caption, we will extract fact types having the same subject and predicate, but verbs which are synonyms (always using the WordNet tool [7]).

– See:

Finally, for each Concept, the number of occurrence will be calculated to determine the preferred representation compared to its synonyms, to put the concept having the high value on the 'See' caption as a preferred word.

Extracting synonyms should be the first operation to do before all extraction, in order to use it to look for information also by synonyms.

'Necessity' & 'Possibility' Captions

A modality will be assigned for each *verb concept*. For this reason, modal verbs such as 'must', 'should', 'always', 'never'…etc. will be extracted from definitional statements (statements with no condition) to be added to '*Verb concepts*' as 'Necessity' or 'Possibility' caption. Modalities at the beginning of statements such as "it is <Modality> that" will be also taken into consideration.

4.5   Output Document

When these steps have been completed, an XML file will be generated containing SBVR based Business Vocabulary in the form of glossary entries.

5   Case Study

In this section, we will illustrate the feasibility of our approach, using following few rules:

R1. All customers must have a credit card
R2. A client can be a moral person
R3. Each car of the rental has one car group
R4. The car's depart is always from California
R5. Moral person is always a company
R6. If the customer holds a credit card, then the customer can rent a vehicle.

Figure 4 illustrate the POS tagging and dependency parsing, using the StanfordNLP tool [3].

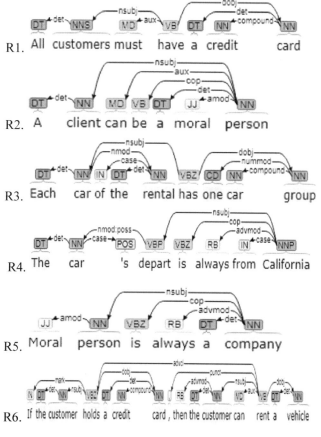

Fig. 4. POS tagging and dependency parsing of statements.

According to the above, we will obtain the following:

- General concept concept:

  {moral_person, person, car_group, car, group, car_depart, depart, car_rental, rental, customer, client, person, company, vehicule, credit_card, credit, card}

- Individual concept:
  {California}

- Verb concept:

  {custmer has credit_card, client is moral_person, car_rental car_group, car_depart is_from California, moral_person is company customer holds credit_card, customer rent vehicule}

Due to the space limit, we will give one example of extracted glossary:
Customer

Definition	: All customers must have a credit card
Definition	: A client can be a moral person
General Concept	: Person
Concept Type	: General concept
Synonym	: Client

For the "Customer" concept found in R1, R2 and R6, we have detected "client" as synonym in R2 which has "Person" as hypernym in R2 what makes it a "General Concept". In the other hand, we have extracted statement having the following pattern "Customer + to be/have + Noun" as a definition. In the same manner, we can observe that the number of occurrence of "Customer" concept is higher than "client" concept, thus it will be added in the "see" caption for "client" concept. Finally, "California" concept is detected as a pronoun, therefore, it will have "Individual concept" as a concept type and "city" as a General Concept.

6   Future Works

Elementary results are encouraging and should be validated by implementing a solution for a larger sample size including complex ones. Further improvements are also needed to remove irrelevant results.

7   Conclusion

In this paper, we have proposed an approach to extract the business vocabulary from natural language business rules, and save it according to the SBVR standard, using Natural Language Processing (NLP). Firstly, POS tagging and dependency parsing, are applied on each statement allowing extraction of SBVR concept, then, information about each concept are extracted and saved in the form of glossary entries. SBVR Structured English (SBVR-SE) notation was chosen – as one of possibly notation mapping to the SBVR Meta-Model due to its popularity. This approach will participate in creating a strong base for Business rules as well as providing a relevant resource for the next software development steps.

References

1. Copyright 2003. Business Rules Group. Version 2.0, 1 November 2003. Edited by Ronald G. Ross. www.BusinessRulesGroup.org
2. Semantics of Business Vocabulary and Rules (SBVR), Version 1.4, Object Management Group (2017). www.omg.org/spec/SBVR/
3. Stanford NLP. nlp.stanford.edu/
4. OMG: Object Management Group. http://www.omg.org/
5. Jackson, D.A.: A Language & Tool for Relational Models (2012). http://alloy.mit.edu/alloy/
6. Object Management Group, Inc.: Unified Modeling Language (UML)
7. Miller, G.A.: WordNet: a lexical database for English. Commun. ACM **38**(11), 39–41 (1995)
8. Bajwa, I.S., Lee, M.G., Bordbar, B.: SBVR business rules generation from natural language specification. Artificial Intelligence for Business Agility—Papers from the AAAI 2011 Spring Symposium (SS-11-03)
9. Bajwa, I.S., Lee, M., Bordbar, B., Ali, A.: Addressing semantic ambiguities in natural language constraints. In: Proceedings of the Twenty-Fifth International Florida Artificial Intelligence Research Society Conference (2012)
10. Bajwa, I.S., Bordbar, B., Lee, M., Anastasakis, K.: NL2 alloy: a tool to generate alloy from NL constraints. J. Digital Inf. Manage. **10**(6), 365–372 (2012)
11. Umber, A., Bajwa, I.S., Asif Naeem, M.: NL-based automated software requirements elicitation and specification. In: Abraham, A., Lloret Mauri, J., Buford, J.F., Suzuki, J., Thampi, S.M. (eds.) Advances in Computing and Communications, ACC 2011. Communications in Computer and Information Science, vol. 191. Springer, Heidelberg (2011)
12. Umber, A., Bajwa, I.S.: A step towards ambiguity less natural language software requirements specifications. IJWA **4**, 12–21 (2012)
13. Umber, A., Bajwa, I.S.: Minimizing ambiguity in natural language software requirements specification. In: 2011 Sixth International Conference on Digital Information Management, 26–28 September 2011
14. Bajwa, I.S., Asif Naeem, M.: On specifying requirements using a semantically controlled representation. In: Muñoz, R., Montoyo, A., Métais, E. (eds.) Natural Language Processing and Information Systems, NLDB 2011. Lecture Notes in Computer Science, vol. 6716. Springer, Heidelberg (2011)
15. Ramzan, S., Bajwa, I.S., Ul Haq, I., Asif Naeem, M.: A model trans-formation from NL to SBVR. In: Ninth International Conference on Digital Information Management (ICDIM 2014), 29 September–1 October 2014
16. Ramzan, S., Bajwa, I.S., Ramzan, B.: A natural language metamodel for generating controlled natural language based requirements. Sci. Int. (Lahore) **28**(3), 2767–2775 (2016). ISSN 1013-5316
17. Roychoudhury, S., Sunkle, S., Kholkar, D., Kulkarni, V.: From natural language to SBVR model authoring using structured English for compliance checking. In: 2017 IEEE 21st International Enterprise Distributed Object Computing Conference

18. Hypsky, R., Kreslikova, J.: Definition of business rules using business vocabulary and semantics. Acta Informatica Pragensia **6**(2), 100–113 (2017). https://doi.org/10.18267/j.aip.103
19. Afreen, H., Bajwa, I.S.: A framework for automated object oriented analysis of natural language software specifications. Int. J. Softw. Eng. Appl. (2012)
20. Thakore, D.M., Patki, R.P.: Extraction of class model from software requirement using transitional SBVR format at analysis phase. Int. J. Adv. Res. Comput. Sci. **3**(7) (2012). ISSN No 0976-5697
21. Mohanan, M., Samuel, P.: Open NLP based refinement of software requirements. Int. J. Comput. Inf. Syst. Ind. Manage. Appl. **8**, 293–300 (2016). ISSN 2150-7988
22. Mohanan, M., Samuel, P.: Software requirement elicitation using natural language processing. In: Snášel,V., Abraham, A., Krömer, P., Pant, M., Muda, A. (eds.) Innovations in Bio-Inspired Computing and Applications. Advances in Intelligent Systems and Computing, vol. 424. Springer, Cham (2016)
23. Njonko, P.B.F., El Abed, W.: From natural language business requirements to executable models via SBVR. In: International Conference on Systems and Informatics (ICSAI2012)
24. Merriam-Webster (2018). Merriam-Webster.com

Arabic Paraphrasing Recognition Based Kernel Function for Measuring the Similarity of Pairs

Hanane Elfaik[1]([⊠]), Mohammed Bekkali[1], Habibi Brahim[2], and Abdelmonaime Lachkar[1]

[1] Laboratory of Engineering Systems and Applications,
Sidi Mohammed Ben Abdellah University, Fez, Morocco
{hanane.elfaik,mohammed.bekkali,abdelmonaime.lachkar}@usmba.ac.ma
[2] Institute of Studies and Research for Arabization,
Mohammed V University, Rabat, Morocco

Abstract. Paraphrasing techniques aim to recognize, generate, or extract linguistic expressions that express the same meaning. These techniques affects positively or negatively the performance of many natural language-processing systems such as Question Answering, Summarization, Text Generation, and Machine Translation.... In this paper, we propose an efficient Arabic paraphrase recognizer based on kernel function and the specificity of terms, which is computed by term co-occurrence and term frequency - inverse document frequency. The experimental results show that our method outperforms the exiting methods based on similarity measures using a standard Arabic paraphrase database PPDB.

Keywords: Paraphrase recognition ·
Natural Language Processing (NLP) · Similarity measure ·
Arabic language

1 Introduction

Paraphrases are pairs of sequences of words, both in the same language, that have the same meaning in at least some contexts. It is the act of generating an alternate sequence of words that conveys the same meaning. Since the meaning of a text is determined only when its context is given.

The identifying paraphrase is an important capability for many natural language processing applications, including machine translation, question answering, summarization and text generation. For example, In Question Answering systems for document collections, a question may be phrased differently than in a document that contains the answer, and considering such variations can improve performance significantly of system.

We can categorize the paraphrase methods in tree types, recognition, generation, or extraction of paraphrases. In this paper, we focus on the paraphrase recognition rather than on the paraphrase generation task and paraphrase

© Springer Nature Switzerland AG 2019
F. Khoukhi et al. (Eds.): AIT2S 2018, LNNS 66, pp. 183–194, 2019.
https://doi.org/10.1007/978-3-030-11914-0_20

extraction. Our task consists in given two text expressions, the system has to determine whether the two texts have the same meaning or not. For example the sentences "وقت قصير" and "قريبا" express the same meaning therefore, they are paraphrases of each other.

Like many other Semitic languages, Arabic is highly inflected; therefore, data sparseness becomes even more noticeable than in English and using paraphrase in the Arabic language turns out to be even more complicated. Arabic words are derived from a root and a pattern (template), combined with prefixes, suffixes. Using the same root with different patterns may yield words with different meanings. Words are inflected for person, number and gender; prefixes and suffixes are then added to indicate definiteness, conjunction, various prepositions and possessive forms. In the case of Arabic, there are very few works that have been proposed, and the paraphrase problem still not solved. With this aim in mind, in this paper we present a new approach for Paraphrase recognition of Arabic language using kernel function for measuring the similarity between two linguistic expressions. The main achievements, including contributions to the field can be summarised as follows: providing a similarity measure to compute the similarity between two sentences taking into consideration the choice of text transformation. The rest of this paper is organized as follows. Section 2 analyses some related work. In Sect. 3, we describe in detail our paraphrase recognizer system. Section 4 provides the experimental study and the results. Finally, we conclude in Sect. 5 and give an outline for future work.

2   Related Work

Paraphrase recognition is concerned with the ability of identifying alternative linguistic expressions of the same meaning at different textual levels (document level, paragraph level, sentence level, word level, or combination between them) [2]. In literature, we can found others labels for Paraphrase Recognition as Paraphrase Detection and Paraphrase Identification [2,3,21]. Available research on the paraphrase recognition approaches can be categorized into supervised and unsupervised learning techniques as follows.

2.1   Supervised Techniques

[10] proposed an approach used MT metrics such as The NIST score [5], Position independent word error rate (PER) [20], Word error rate (WER) [19], The BLEU score [17]. The text pre-processing task consisted of tokenization, POS tagging and stemming. They used also semantic similarity distance measure computed based on WordNet-based lexical relationship measures [12]. This approach achieved the best performance when using support vector machine (SVM) and a combination of all measures with accuracy of 74.96% on Microsoft Research Paraphrase Corpora [6].

[13] measures lexical and semantic similarity with the combination of three machine-learning classifiers (SVM, K-Nearest Neighbor (K-NN) and Maximum

Entropy (MaxEnt). Semantic similarity features were based on WordNet. The experiments showed that, the best performance was achieved by the lexical feature set comparing with the similarity feature, Moreover, the performance enhanced by 1% when combined the two features sets. The best result between all classifier obtained with SVM.

[21] provided an approach using semantic heuristic features to identify the paraphrasing. The text-pre-processing task consisted of tokenization, POS tagging and stop words removing. They also used monotonic and no-monotonic alignment and semantic heuristics to define feature vectors' set, then they performed machine learning techniques along with these vectors using Weka tool to predict PI. The approach improving the accuracy compared to state-of-the-art PI systems.

[7] proposed an approach using features based on overlap of word and character n-grams and train support vector machine (SVM). Their results demonstrate that character and word level overlap features in combination can give performance comparable to methods employing more sophisticated NLP processing tools and external resources.

[1] proposed a state-of-the-art approach for paraphrase identification and semantic text similarity analysis in Arabic news tweets. They trained two machine-learning classifiers (Maxi-mum Entropy (MaxEnt) and Support Vector Regression (SVR)) on features' vectors extracted from Lexical, syntactic, and semantic attributes. The experimentation results show that the approach achieves good results in comparison to the baseline results on a dataset prepared for this research.

2.2   Unsupervised Techniques

[15] presents a method for measuring the semantic similarity of texts, using corpus-based and knowledge-based measures of similarity. They combined two types of corpus-based measures, pointwise mutual information (PMI) [14] and latent semantic analysis (LSA) [4], and six types of knowledge-based. This approach achieved a good performance with accuracy of 71.5% on the MRP corpora with a threshold prediction value of 0.5.

[9] present an algorithm for paraphrase identification which makes extensive use of word similarity information derived from WordNet [8]. The approach achieved good performance with accuracy of 74% on MSR corpora with a threshold of 0.8 for the similarity decision.

[11] proposed a method for measuring the semantic relatedness of words producing a Salient Semantic Analysis (SSA) model for this task. They used semantic profiles from salient encyclopaedic features taken from encyclopaedic knowledge. This model is built on the notion that the meaning of a word can be characterized by the salient concepts found in its immediate context. It has outstanding performance in comparison with corpus-based and knowledge-based semantic relatedness models.

[16] proposed a comparative study between neural word representations and traditional vector spaces based on co-occurrence counts, in a number of

compositional tasks. They use three different semantic spaces and implement seven tensor-based compositional models, which they then test in tasks involving verb disambiguation and sentence similarity. The results show that on the small-scale tasks, the neural vectors gave a result better than or similar to the count based vectors, whereas on the large-scale tasks the neural word embedding gave a result better than the co-occurrence based.

3   Architecture

In this section, we describe the details of our proposed architecture for Paraphrase recognition. Our proposed architecture consists of three main components: Pre-processing, Measuring Similarity and Paraphrase judgement. The system architecture is presented in Fig. 1.

3.1   Pre-processing

Sentences pre-processing, which is the first step in our approach, converts the Arabic sentences to a form that is suitable for our method of similarity measure. These pre-processing tasks include Punctuations removal, Latin characters removal, Digits removal, Tokenization and normalization. These linguistic are used to reduce the ambiguity of words to increase the accuracy and the effectiveness of our approach. The pre-processing of Arabic sentences consists of the following steps:

- Tokenization:

Tokenization is a method for dividing texts into tokens; Words are often separated from each other by blanks (white space, semicolons, commas, quotes, and periods). These tokens could be individual words (noun, verb, pronoun, and article, conjunction, preposition, punctuation, numbers, and alphanumeric) that are converted without understanding their meaning or relationships. The list of tokens becomes input for further processing.

- Stop Word Removal:

Stop word removal involves elimination of insignificant words, such as (since/منذ) (for/لأجل) and (so/لذا), which appear in the sentences and do not have any meaning or indications about the content.

   Other examples of these insignificant words are articles, conjunctions, pronouns (such as he/هو, she/هي, and they/هم), prepositions (such as from/من, to/الى, in/في, and about/حول), demonstratives, (such as this/هذا, these/هؤلاء, and there/اولئك), and interrogatives (such as where/أين, when/متى, and whom/لمن). Moreover, Arabic circumstantial nouns indicting time and place (such as after/بعد, above/فوق, and beside/بجانب), signal words (such as first/اولا, second/ثانيا, and third/ثالثا). A list of 202 words was prepared to be eliminated from all the sentences.

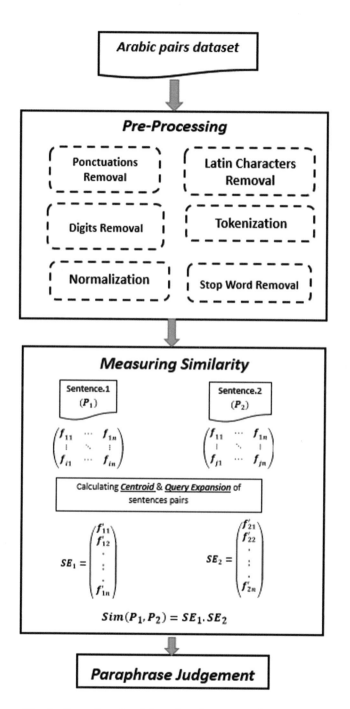

Fig. 1. Overall methodology for the proposed approach

- Punctuations Removal:

Punctuations Removal aims to remove the punctuations symbols such as $\{\#, -, -, ., ., , ; , :, ,\}$, because these symbols are not useful in our approach.

- Latin Characters Removal and Digits Removal:

In addition, we remove the Latin characters and digits, which appear in the sentences and do not have any meaning or indications in our method.

- Word Normalization:

Normalization aims to normalize certain letters that have different forms in the same word to one form. For example, the normalization of " ء " (Hamza), " آ " (aleph mad), " أ " (aleph with hamza on top), " ؤ " (hamza on waw), " إ " (aleph with Hamza at the bottom), and " ئ " (Hamza on ya) to " ا " (aleph). Another example is the normalization of the letter " ى " to " ي " and the letter " ة " to " ه ".
We remove the diacritics such as }ﹰ ﹴ ﹺ ﹸ ﹶ ﹾ{ because these diacritics are not used in extracting the Arabic roots and not useful in the approach proposed. Finally, we duplicate the letters that include the symbol "الشدة ّ " , because these letters are used to extract the Arabic roots and removing them affects the meaning of the words.

3.2 Measuring Similarity

The following section provides details of our kernel function for measuring similarity between two sentences where each sentence P_i is represented by a vector of terms t_j^i, given by:

$$P_i \left(t_1^i \ t_2^i \ \cdots \ t_n^i \right)$$

where n is the number of terms in sentence i.

We take into account the specificity of terms in the sentence, it is determined using:

- Term co-occurrence :

The feature vector of a given sentence is formed by taking into account the occurrence of terms in the corresponding sentence.

- Term frequency – inverse document frequency of term:

TFIDF is evolved from IDF which is proposed by [18] with heuristic intuition that a query term which occurs in many documents is not a good discriminator, and should be given less weight than one which occurs in few documents. Equation 1 is the classical formula of TFIDF used for term weighting.

$$w_{ij} = t_{ij} * \log \left(\frac{N}{df_i} \right) \tag{1}$$

Where w_{ij} is the weight for term i in sentence j, N is the number of sentence in the collection, t_{ij} is the term frequency of term i in sentence j and df_i is the document frequency of term i in the collection.

We adopt several phases to obtain the kernel similarity between two sentences in the corpus by applying the following steps:

First, for each term t_j^i in the sentence P_i where n is the number of terms in sentence i, we try to identify its specificity. After the first step we have obtain

$$V_i \left(v_1^i \; v_2^i \; \cdots \; v_n^i \right)$$

where $v_j^i \in \mathbb{R}^k$ and k is the number of sentences in corpus i.

Second, normalized the serval vectors of terms in the sentence by the following formula:

$$Normalized(v_j^i) = \frac{v_j^i}{\left\| v_j^i \right\|} \qquad (2)$$

Third, we calculate the centroid of term's vector, given by:

$$C(V_i) = \frac{1}{n} \sum_{j=1}^{k} Normalized(v_j^i) \qquad (3)$$

Forth, we calculate the sentence expansion by the following formula:

$$SE_i = \frac{C(V_i)}{\left\| C(V_i) \right\|} \qquad (4)$$

Finally, given that, we have a means for computing the sentence expansion; it is a simple matter to define the similarity as the inner product of the sentence expansion. More formally, given two sentence P_i and P_j, we define the similarity between them as :

$$K(P_i, P_j) = SE_i \cdot SE_j \qquad (5)$$

This similarity score has a value between 0 and 1, with a score of 1 indicating identical text segments, and a score of 0 indicating no semantic overlap between the two segments.

4   Experiments and Results

4.1   Paraphrasing Corpora

Experimenting with paraphrase recognizers requires datasets containing both positive and negative input pairs. The most widely used benchmark dataset for paraphrase recognition in English language is the Microsoft Research (MSR) Paraphrase Corpus. Unfortunately, in the case of Arabic language, there are very few corpora interested to this problem. We have collected our dataset from PPDB database, the resource is derived from large volumes of bilingual parallel data. It includes a collection of paraphrases for 21 additional languages:

Arabic, Bulgarian, Chinese, Czech, Dutch, Estonian, Finnish, French, German, Greek, Hungarian, Italian, Latvian, Lithuanian, Polish, Portuguese, Romanian, Russian, Slovak, Slovenian, and Swedish. The multilingual paraphrase database is freely available. Then the dataset was annotated manually by experts. These later were asked to assign a binary values (0 = "False"/1 = "True") to the pairs of language expressions indicating whether the pair are paraphrase or not.

Over 1408943 Arabic pairs selected from PPDB, However, only 6806 sentences were manually annotated, 5902(86.7%) true paraphrase pairs whereas 904(13.3%) false paraphrases.

4.2   String Similarity Measures

We compare the results of the proposed method with five existing string similarity measures. Given two input text segment (p, q); the idea of surface-matching methods is based on the number of words that occur in both text segments. Suppose P and Q are the sets of words in p and q, respectively. Common similarity measures discussed in [14] are listed:

$$Matching = |P \cap Q| \tag{6}$$

$$Dice = \frac{2|P \cap Q|}{|P| + |Q|} \tag{7}$$

$$Jaccard = \frac{|P \cap Q|}{|P \cup Q|} \tag{8}$$

$$Overlap = \frac{|P \cap Q|}{\min(|P|, |Q|)} \tag{9}$$

$$Cosine = \frac{|P \cap Q|}{\sqrt{|P| * |Q|}} \tag{10}$$

Table 1. Proposed approach results compared with the existing similarity measures

	Accuracy	Precision	Recall	F-measure
Cosine	0.55	0.49	0.55	0.48
Jaccard	0.52	0.48	0.52	0.45
Dice	0.55	0.48	0.55	0.48
Overlap	0.54	0.49	0.54	0.47
Matching	0.41	0.48	0.42	0.34
Our Method co-occurrence	0.62	0.58	0.62	0.56
Our Method TFIDF	0.80	0.85	0.81	0.83

4.3   Performance Measure

In this section, we define performance measures for classification, which are widely used in the information retrieval community and machine learning. Four commonly used evaluation measures are used: accuracy, precision, recall, and F-measure with equal weight on precision and recall. These measures are defined below. TP (true positives) and FP (false positives) are the numbers of pairs that have been correctly or incorrectly, respectively, classified as positive (paraphrases). TN (true negatives) and FN (false negatives) are the numbers of pairs that have been correctly or incorrectly, respectively, classified as negative (not paraphrases).

$$accuracy = \frac{TP + TN}{TP + TN + FP + FN} \tag{11}$$

$$precision = \frac{TP}{TP + FP} \tag{12}$$

$$recall = \frac{TP}{TP + FN} \tag{13}$$

$$F - measure = \frac{2 * precision}{precision + recall} \tag{14}$$

4.4   Results

In this section, we present the results obtained by the Arabic paraphrase recognition. We evaluate the results in terms of accuracy, representing the number of correctly identified true or false classifications on the test data set. We also measure precision, recall and F1-measure, calculated with respect to the true values in the test data. Table 1 shows the results obtained.

Table 1 show that our approach both with co-occurrence and with TFIDF outperforms for all five of the similarity measures used in these experiments. In addition, TFIDF shows best results compared to term co-occurrence.

4.5   Discussion

Using four performance measures namely accuracy, precision, Recall and F1 measure calculated we can clearly see the performance of our approach with both co-occurrence and TFIDF. The best results quoted in Table 1 were found using the threshold '0.40' for deciding whether two-language expressions are paraphrase or not. We assign a binary values of (0 = "False"/1 = "True") to the pairs of language expressions compared with the manually annotated decision. Figure 2 shows the accuracy with various thresholds.

As discussed above, when the threshold is '0.40' we obtain the best accuracy in both TFIDF transformation and co-occurrence transformation, and this is the raison behind the choice of this value as a threshold of decision. The experiments also show that the choice of the text transformation influence on the results.

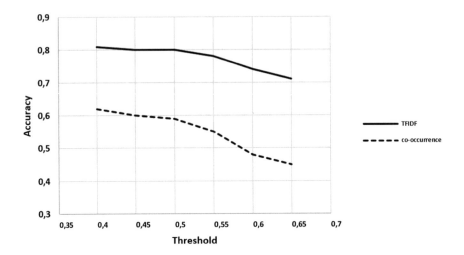

Fig. 2. The accuracy with various thresholds

Using TFIDF transformation of term outperforms the transformation based on co-occurrence of term in the document and this due to that the TFIDF measure combines two aspects of a word: the importance of a word for a document and its discriminative power within the whole collection. These two aspects match well with the general intuition for suitability of words as good keywords.

5   Conclusion

In this paper, we proposed a system for Arabic paraphrasing recognition using an efficient method to recognize whether two language expressions convey the same meaning. Our approach is based on kernel function to measure the similarity between two sentences. The system was evaluated on a dataset prepared for this research with a combination with experts of Arabic language.

A series experiments results prove that the performance of our approach outperforms five existing similarity measures. The experiments also show that the choice of the text transformation influence on the results. Indeed, the experiments have shown that the performance by TFIDF weighing produce better results more than co-occurrence.

However, In Question answering system, a question may be phrased differently than in a document that contains the answer, and taking such variations into account can improve performance significantly of system. In our future work, we plan to incorporate the results of paraphrasing recognition to achieve an automatic question answering system.

References

1. Al-Smadi, M., Jaradat, Z., Al-Ayyoub, M., Jararweh, Y.: Paraphrase identification and semantic text similarity analysis in arabic news tweets using lexical, syntactic, and semantic features. Inf. Process. Manage. Int. J. **53**(3), 640–652 (2017)
2. Bhagat, R., Hovy, E.: What is a paraphrase? Comput. Linguist. **39**(3), 463–472 (2013)
3. Bhagat, R., Hovy, E., Patwardhan, S.: Acquiring paraphrases from text corpora. In: Proceedings of the Fifth International Conference on Knowledge Capture, pp. 161–168. ACM (2009)
4. Dennis, S., Landauer, T., Kintsch, W., Quesada, J.: Introduction to latent semantic analysis. In: Slides from the Tutorial given at the 25th Annual Meeting of the Cognitive Science Society, Boston (2003)
5. Doddington, G.: Automatic evaluation of machine translation quality using n-gram co-occurrence statistics. In: Proceedings of the Second International Conference on Human Language Technology Research, pp. 138–145. Morgan Kaufmann Publishers Inc. (2002)
6. Dolan, B., Quirk, C., Brockett, C.: Unsupervised construction of large paraphrase corpora: exploiting massively parallel news sources. In: Proceedings of the 20th International Conference on Computational Linguistics, p. 350. Association for Computational Linguistics (2004)
7. Eyecioglu, A., Keller, B.: Twitter paraphrase identification with simple overlap features and svms. In: Proceedings of the 9th International Workshop on Semantic Evaluation (SemEval 2015), pp. 64–69 (2015)
8. Fellbaum, C.: A semantic network of english verbs. WordNet: Electron. Lexical database **3**, 153–178 (1998)
9. Fernando, S., Stevenson, M.: A semantic similarity approach to paraphrase detection. In: Proceedings of the 11th Annual Research Colloquium of the UK Special Interest Group for Computational Linguistics, pp. 45–52 (2008)
10. Finch, A., Hwang, Y.S., Sumita, E.: Using machine translation evaluation techniques to determine sentence-level semantic equivalence. In: Proceedings of the Third International Workshop on Paraphrasing (IWP2005) (2005)
11. Hassan, S., Mihalcea, R.: Semantic Relatedness using Salient Semantic Analysis. AAAI press, San Francisco (2011)
12. Jiang, J.J., Conrath, D.W.: Semantic similarity based on corpus statistics and lexical taxonomy. arXiv preprint cmp-lg/9709008 (1997)
13. Kozareva, Z., Montoyo, A.: Paraphrase identification on the basis of supervised machine learning techniques. In: Advances in Natural Language Processing, pp. 524–533. Springer (2006)
14. Manning, C.D., Manning, C.D., Schütze, H.: Foundations of Statistical Natural Language Processing. MIT press (1999)
15. Mihalcea, R., Corley, C., Strapparava, C., et al.: Corpus-based and knowledge-based measures of text semantic similarity. In: AAAI, vol. 6, pp. 775–780 (2006)
16. Milajevs, D., Kartsaklis, D., Sadrzadeh, M., Purver, M.: Evaluating neural word representations in tensor-based compositional settings. arXiv preprint arXiv:1408.6179 (2014)
17. Papineni, K., Roukos, S., Ward, T., Zhu, W.J.: Bleu: a method for automatic evaluation of machine translation. In: Proceedings of the 40th Annual Meeting on Association for Computational Linguistics, pp. 311–318. Association for Computational Linguistics (2002)

18. Spärck Jones, K.: IDF term weighting and IR research lessons. J. Documentation **60**(5), 521–523 (2004)
19. Su, K.Y., Wu, M.W., Chang, J.S.: A new quantitative quality measure for machine translation systems. In: Proceedings of the 14th Conference on Computational Linguistics, vol. 2, pp. 433–439. Association for Computational Linguistics (1992)
20. Tillmann, C., Vogel, S., Ney, H., Zubiaga, A., Sawaf, H.: Accelerated DP based search for statistical translation. In: Fifth European Conference on Speech Communication and Technology (1997)
21. Ul-Qayyum, Z., Altaf, W.: Paraphrase identification using semantic heuristic features. Res. J. Appl. Sci. Eng. Technol. **4**(22), 4894–4904 (2012)

Model of the Structural Representation of Textual Information and Its Method of Thematic Analysis

Faddoul Khoukhi[✉]

Faculty of Sciences and Techniques Mohammedia, Mohammedia Computer
Science Laboratory (LIM), BP 146, 20650 Mohammedia, Morocco
khoukhif@gmail.com

Abstract. In the information retrieval system, often the user making the request, has a rough idea about the topic of his document to find, but the system's response is not always satisfactory. This is probably due to incorrect training of the user's request. The problem of the constitution of a correct query of the information can be justified by causes such as the absence of knowledge on the keywords which determine the document to be found, the lack of experience and qualification for the construction of such queries and the absence of stable and usable terminologies in the field in question. In this work, our goal is to obtain the adequate information desired by the user. Among the variants for the resolution of this problem is the search for documents by structure. The user in this case associates a given document with a structure, whereas the system performing this search variant must select the appropriate documents (in content and in theme).

Keywords: Thematic analysis · Structural representation of information · Information flow · Semantic relation · Semantic similarity

1 Introduction

The thematic analysis methods of unstructured textual documents are intended for the resolution of document search problems by structure. In this paper, we present a representation model of textual documents in the form of a multigraph oriented structure.

In this paper, we present a representation model of textual documents in the form of a multigraph oriented structure. To do this, two main problems are posed:

- The thematic extraction of textual information;
- Thematic similarity calculation of documents.

1.1 Thematic Extraction of the Document

The theme reflects the content of the document and contains a set of keywords with some correspondence between them. Among the variants of these correspondences we

F. Khoukhi et al. (Eds.): AIT2S 2018, LNNS 66, pp. 195–203, 2019.
https://doi.org/10.1007/978-3-030-11914-0_21

find the coefficients of the weights which give information on the value of such a word in a concrete theme.

1.2   Thematic Similarity Calculation of Documents

It is the result of the calculation of the thematic similarity that determines the result of the search for the information. Often the result of the search [1, 2] gives a set of documents that more or less satisfy the search criteria.

The analysis of the existing methods carrying out search systems of documents by structure [1–3] made it possible to note a insufficiency of adequate solutions in this field because of absence of enough theories and practices for the resolution of the problem of unstructured thematic analysis of the natural language of textual documents.

Solving the problem of thematic analysis is more in demand not only in the field of research systems but generally in all systems of analysis and information processing. It is a broad spectrum of various problems of intellectual information processing such as the extraction, identification and recognition of the meaning of the content of a dialogue. All this makes the research work in the field of thematic analysis and the processing of unstructured information very timely and useful.

2   Méthods of Thematic Analysis

Despite the various methods of thematic analysis that exist up to now, such as linguistic analysis [10], statistical analysis [11] and latent semantic analysis (LSA) [12]. The problem of adequately representing the information needs of users remains an open problem.

In this paper, we use an approach that is based on the representation of the user query in the form of a document-model and the realization of an effective thematic analysis method that must identify exactly the theme of the documents.

3   Modeling Text as a Multigraph

In this paper, the problem of the thematic classification of text is solved by modeling the text using a flow of information oriented in the form of a multi-graph whose vertices and edges represent respectively the words and the relations between them. This multigraph is an informational structure of the text. By the informational structure, we thus designate the set of all the words constituting the text and the relations between them. To obtain the structural representation, the text is considered as a flow of information represented by informative elements (words).

If we take successively the words of the text starting from the first to the last then we form the information flow F. By grouping the unique words constituting the texts (without repetition) we form the set of informative elements $I : I = \{i_1, i_2, \ldots, i_n\}$ where i is an informative element corresponding to a given word of the text. Information flow F will be represented with a series of these elements (Fig. 1).

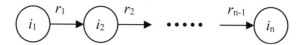

Fig. 1. The flow of information

where: $r_i = (i_i, i_{i+1})$ is the relation between two successive informative elements in the stream. $F = (i_1, i_2, \ldots, i_n)$.

The order of alternation of the informative elements depends on their positioning in the text. The informative elements can be repeated in the text. An important condition is that an informative element matches one and a single word of text. The identical words of the text will therefore be represented by the same information element.

Example:

That is the following fragment of the text: "By combining a few words we form phases. By combining a few letters we form words: The set of computer elements corresponding to this fragment of the text is: $I = \{i_1, i_2, i_3, i_4, i_5, i_6, i_7, i_8, \ i_9\}$, where

$i_1 = $ combinant, $i_2 = $ des, $i_3 = $ en, $i_4 = $ forme, $i_5 = $ lettres, $i_6 = $ mots, $i_7 = $ on, $i_8 = $ phrases, $i_9 = $ quelques. The information flow corresponding to this text fragment (Fig. 2) will be:

$$F = (i_3, i_1, i_9, i_6, i_7, i_4, i_2, i_8, i_3, i_1, i_9, i_5, i_7, i_4, i_2, i_6).$$

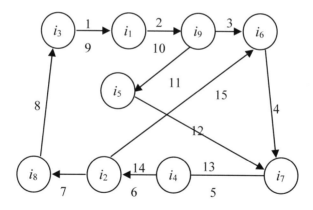

Fig. 2. The structure formed by the information flow

The flux obtained thus makes it possible to form the information structure schematized in Fig. 2.

The repetition of the words in the text indicates that the flow of information will pass several times through the same information element.

The indexing of the flow consists of assigning an index to each passage between two informative elements that are incremented step by step. From the information structure we extract some properties particularities of the flow of information which will allow to elaborate the algorithm for the thematic classification of the text.

We therefore designate by n (I) $= |$ I $|$ the number of informative elements of set I (number of unique words in the text), and by n (F) $= |$ F $|$ the number of informative elements constituting the flow of information F (the total number of the words of the text). M(I, R) represents the information structure (multigraph oriented) where I is a set of informative elements and $R = (r_1, r_2, \ldots, r_{n-1})$ is the set of all relationships between these elements. The set R may contain some repetitions of the relations between the informative elements in the case where the flow F of information passes several times through the same information element of the text.

$$\forall i \in I \exists R(i) = ((r, r_{i+1})_1, \ldots, (r_j, r_{j+1})_n)$$

For each information element belonging to I of the structure M(I, R), there exists a set of n pairs r_i, r_j et r_{i+1}, r_{j+1} respectively denoting input and output relations (relations with previous and next words).

We denote by n(R(i)) the number of relation pairs in R(i) which characterizes the number of relations of a given information element with the other elements of the structure M(I, R). It therefore corresponds to the number of repetitions of the word in the text.

The number of relationship pairs is then written:

n(R(i)) = d(i), where d(i) represents the degree of importance of an informative element: $\forall i \in I$, $0 < d(i) \leq n(I) - 1$.

We denote by d(M(I, R)) max and d(M(I, R)) min respectively the maximum and minimum degree of the informative element for the information structure M(I, R):

$$d(M(I, R))_{max} = \max d(i) \text{ and } d(M(I, R))_{min} = \min d(i) \quad i \in I.$$

$r^+(F, i_k, i_j, e)$ et $r^-(F, i_k, i_j, e)$ represent the distance between two informative elements in flow F, where:

- e is the order of the information element ik in the stream F (its position number in F).
- r^+ and r^- mean that the distance is measured respectively in the direct and inverse direction of the flow of information. The distance is evaluated in the number of information elements existing between i_k and i_j add 1.

Example, the next flow of information:

$$F = (i_2, i_5, i_1, i_4, i_5, i_2, i_3)$$

$$r^+(F, i_2, i_3, 1) = 6, r^+(F, i_2, i_3, 2) = 1, r^-(F, i_1, i_2, 1) = 2, \ r^-(F, i_1, i_3, 1)$$
$$= 0, \ r^+(F, i_1, i_3, 1) = 4.$$

0 means that the distance is indeterminate.

$r_{min}(F, i_k, i_j)$ represents the minimum distance between two pieces of information in the stream.

Example:

$$F = (i_2, i_6, i_1, i_4, i_5, i_7, i_2) \, r_{\min}(F, i_2, i_5) = 2, r_{\min}(F, i_1, i_2) = 2, r_{\min}(F, i_1, i_8) = 0.$$

We designater $F(i, e, [r^-, r^+])$ the flow of information relative to an information element i,

where e is the order of the information element i_k in stream F and $[r^-, r^+]$ is a neighborhood for which the flow of information is defined.

$$F(i, e, [r^-, r^+]) = (i - r^-, \ldots, i - 2, i - 1, i, i + 1, i + 2, \ldots, i + r^+),$$

where $i \pm n$ is the index of an informative element in the flow F with respect to the information element i, $i + 1$ et $i - 1$ represent respectively the informative element that comes right after and before the element i in the flow F.

Example:

Either the flow of information defined by: $F = (i_2, i_5, i_1, i_4, i_5, i_2, i_3, i_8, i_{10}, i_1)$

$$F(i_4, 1, [3, 3]) = (i_2, i_5, i_1, i_4, i_5, i_2, i_3). \, F(i_2, 2, [2, 4]) = (i_4, i_5, i_2, i_3, i_8, i_{10}, i_1).$$

$$F(i_2, 1, [2, 2]) = (0, 0, i_2, i_5, i_1), F(i_5, 1, [2, 5]) = (0, i_2, i_5, i_1, i_4, 0, 0, 0).$$

In this case, 0 is assigned to all items outside the designated area. The set of all relative information flows with respect to information element i in flow F:

$$D(F, i, [r^-, r^+]) = \bigcup_{e=0}^{d(i)} F_e(i, e, r^-, r^+),$$

Example: Either the next flow:

$F = (i_2, i_5, i_1, i_4, i_5, i_2, i_3, i_8, i_5, i_1), d(i_5) = 3,$ therefore $D(F, i_5, [2, 2]) =$
$((0, i_2, i_5, i_1, i_4), (i_1, i_4, i_5, i_2, i_3), (i_3, i_8, i_5, i_1, 0)).$

4 Proposed Approach

The graphical modeling of the text obtained above made it possible to illustrate a semantic relation of its content in the form of an information structure and to highlight some relations and regularities in order to approach a method of classification of its thematic content. For the resolution two main problems:

– Thematic extraction of textual information;
– Calculating the degree of thematic belonging of the text to a given class.

Most of the approaches to the thematic classification of texts [4–8], are based on the following hypothesis: the vocabulary and the frequency of use of words depend on the subject of the text [6].

According to the structural representation of the text proposed in this article, the keywords will therefore be the nodes of the graph through which the flow passes several times and the information elements that correspond to them have more relationship with the other informative elements of the graph. The problem remains to automatically determine the threshold that distinguishes keywords from other words in the text.

It is obvious that the choice of the threshold depends on a concrete text, so it depends on the characteristics of the graphical model as d(M(I, R)) max, d(M(I, R)) min and n(I). However this is not enough to determine the theme of the text. The adequacy of the themes (represented by the machine) in relation to those defined by the person is until now an open problem. In this present work, to have a correct and adequate representation of the theme of the text, it is proposed to complete the key words of the text by their contexts. For this we proceed with the following steps:

1. Text modeling and information structure formation M(I, R).
2. The extraction of the set of all the informative elements arranged according to their degrees d(i) (the number of repetitions in the text). The elements with d(M(I, R)) max will be the first and so on.
3. Extracting all keywords.

From the informative elements (obtained in the preceding step) one selects the first n (n is determined according to the value of the threshold chosen), which will represent the primary group of the key words $S_p = \{k_1 i_1, k_2 i_2, \ldots, k_n i_n\}$. The coefficients $k_1, k_2, \ldots, k_n$ correspond to the degrees of the informative elements.

4. Formation of the secondary group S_s which gives more contextual precision based on the contextual analysis of the information elements of S_p.
5. Get the global set that determines the theme of the text: $S = S_p + S_s$.

The idea of the method is to form the primary group of key words based on the number of informative elements and their repetitions in the text and then supplement them with their context using the characteristics properties of the multigraph. The surroundings of the informative element, set of elements containing in $D(F, i, [r^-, r^+])$, represent the context.. The contextual analysis therefore consists in analyzing the surroundings of a given information element.

For the formation of the set Ss based on the contextual analysis of the informative elements of the set Sp one proceeds as follows:

(1) we denote by $A(i)$ the set of all the flows passing through each information element in the neighborhood in a given neighborhood r. $A(i) = D(F, i, [r^-, r^+])$, with $c\, r^- = r, r^+ = r$. $A(i) = D(F, i, [r, r])$.

We then gather all the A(i), for each element $i \in S_p$ in a common set $A(S_p)$:

$$A(S_p) = \bigcup_{k=0}^{n} A(i_k), i_k \in S_p, 0 \leq k \leq n(S_p)$$

where we designate $A : A = A(S_p)$.

In the final result we obtain a global set A containing all the flows that pass through the information elements of the set Sp in the neighborhood of r.

(2) We extract from A all the information elements belonging to S_p:

$$A \rightarrow A',$$

$$\forall i \in A \Rightarrow (i \in A' \Leftrightarrow i \notin S_p).$$

(3) At the end from the set A' the secondary group of the information elements S_s s constructed by calculating the number of repetitions k_i of each informative element in $A' : \mathbf{A'} \rightarrow \mathbf{S_s}$.

$$S_s = \{k_1 i_1, k_2 i_2, \ldots, k_n i_n\},$$

where the coefficients $k_1, k_2, \ldots, k_n$ represent the repetition number of the informative elements of A.

All the elements of the set S_s exist in some surroundings elements of the set S_p, and each informative element $i \in S_p$ determines the center of certain surroundings in the structure M(I, R).

In Fig. 3 is illustrated an example of selection of a neighborhood of an information element i_5 of an information structure for $r = 1$.

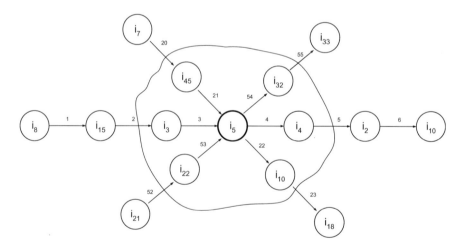

Fig. 3. Selection of around an information element

For this example $S_s = \{i_{45}, i_3, i_{22}, i_{10}, i_4, i_{32}\}$, all the coefficients are equal to 1.

In the case of a textual document, for a given word all its locations in the text are determined. Then for each of its locations we determine the neighboring words that are related to the chosen word for a given distance r (in the forward and reverse direction). Subsequently, the number of repetitions of all the words obtained in the vicinity of the locations is calculated. The repetition numbers thus make it possible to determine the weights of these words in the theme.

In the end we get the overall set of keywords that determine the theme of the text:

$$S = S_p + S_s, S = \{k_1 i_1, k_2 i_2, \ldots, k_n i_n\}.$$

The coefficients $k_1, k_2, \ldots, k_n$, determine the importance of the informative elements in the theme of the text. It should be noted that this method of extracting the theme depends on two parameters such as the value of the threshold and the distance r which respectively determine the key word and its surroundings. The variation of these two parameters can influence considerably the result of the extraction of the theme. For this, we can confirm that the choice and the adequate evaluation of these parameters depend on the calculation of the thematic belonging of the text and its estimation by an expert.

5   Algorithm for Calculating the Thematic Degree of Belonging of a Text to a Structure

Let $S = \{k_1 i_1, k_2 i_2, \ldots, k_n i_n\}$ be a set of key elements of a text structure: where the coefficients $k_1, k_2, \ldots, k_n$ represent the weights of the informative elements and defining the importance of a given element in the theme of the model text.

Let $S_f = \{k_{f1} i_1, k_{f2} i_2, \ldots, k_{fn} i_n\}$ be a set of keywords obtained by the search result of a text (document found by a search system), which we must analyze its thematic similarity with respect to the model text. For this, the information elements $i_1, i_2, \ldots, i_n$ determining the theme of the two sets S and S_f relate to each other, that is to say the elements i_1 of S and S_f are identical. However this does not mean that all the informative elements of S are obligatorily in S_f, a part of them can not exist if such keywords do not belong to the text to analyze.

Traditionally in information retrieval systems that are based on the vector model, for calculating the similarity between vectors, the cosine between them is measured [9]. In our case we will not apply this method because the cosine determines the measure of the proximity of the vectors and not their components. The thematic proximity ω_i according to each of the informative elements will be calculated as the difference of the coefficients of the weights:: $\omega_i = |k_i - k_{fi}|$.

The overall coefficient of thematic proximity for the entire text will be calculated as the sum of all ω_i : $\omega = \sum_{i=1}^{n} \omega_i$.

The calculation of the coefficient of proximity ω for each document found (text) allows to classify the documents according to their thematic similarity.

6   Conclusion

The model developed for the structural representation of textual information, the method and the algorithm of its thematic analysis make it possible to carry out the thematic classification and to calculate the thematic degree of belonging of a text to a structure. The proposed model, the method and the algorithm can be used for solving the concrete problems of document search by structure as well as for the resolution of problems in general such as the analysis and the processing of information.

References

1. Béchet, N., Roche, N., Chauché, J.: ExpLSA et classification of texts. Équipe TAL, LIRMM - UMR 5506, CNRS, Université Montpellier – France
2. Kessler, R., Torres-Moreno, J.M., El-Beze, M.: Text classificationTypical email classification with supervised, semi supervised and unsupervised learning, Université d'Avignon, Laboratoire d'Informatique d'Avignon, France
3. Mokrane, A., Dray, G., Poncelet, P.: Thematic feature of collection of textual documents. Information Systems Day elaborated, île Rousse 2005, Nîmes, France
4. Chakrabarti, S., Berg, M., Dom, B.: Focused crawling: a new approach to topic-specific web resource discovery. In: Proceedings of the WWW8, May 1999
5. Hatzivassiloglou, V., Gravano, L., Maganti, A.: An investigation of linguistic features and clustering algorithms for topical document clustering. In: Proceedings of the SIGIR 2000 (2000)
6. Singhal, A., Mitra, M., Buckley, C.: Learning routing queries in a query zone. In: Proceedings of the SIGIR 1997, pp. 25–32, July 1997
7. Salton, G., Allan, J., Singhal, A.: Automatic text decomposition and structuring. Inf. Process. Manag. 32(2), 127–138 (1996)
8. Salton, G., Singhal, A., Mitra, M., Buckley, C.: Automatic text decomposition and summarization. Inf. Process. Manag. 33(2), 193–208 (1997)
9. Lelu, A.: Comparison of three similarities used in automatic documentations and textual analysis. In: 6th International Days of Statistical Analysis of Textual Data, University of Franche-Comté, Besançon, France
10. Declerck, T.: DFKI GmbH, Language Technology Lab Piroska Lendvai, Research Institute for Linguistics, Department of Language Technology, Hungarian Academy of Sciences
11. Turenne, N.: Statistical learning for extracting concepts from texts. Application to the filtering of textual information. Thesis defended at the University Louis-Pasteur Strasbourg U.F.R Mathematics-Computer Science, ENSAIS de Strasbourg

IR Big Data, IA and Knowledge Management

Plagiarism Detection in Across Less Related Languages (English-Arabic): A Comparative Study

Hanane Ezzikouri[1(✉)], Mohammed Erritali[2],
and Mohamed Oukessou[1]

[1] Applied Mathematics and Scientific Computing Laboratory (LMACS),
Faculty of Sciences and Techniques, Sultan Moulay Slimane University,
BP: 523, Beni-Mellal, Morocco
ezzikourihanane@gmail.com
[2] TIAD Laboratory, Computer Sciences Department,
Faculty of Sciences and Techniques, Sultan Moulay Slimane University,
BP: 523, Beni-Mellal, Morocco

Abstract. Plagiarism is the reuse of someone else's ideas, results, or words without acknowledging the original source. The Plagiarism does not confess linguistic boundaries. Cross-Language (CL-Plagiarism) or multilingual plagiarism occurs if an excerpt written in a different language is translated and no reference of the original source is given. Ranging from translating and changing texts into semantically equivalent to adopting ideas, without giving credit to its originator, CL-Plagiarism can be of many different natures.

In this paper, we compare three recently proposed cross-language plagiarism detection methods Cross Language Character N-Gram (CL-CNG), based on language syntax, Cross Language Alignment Based Similarity Analysis (CL-ASA) based on statistical translation, and Translation plus monolingual analysis (T+MA) based on machine translation and monolingual similarity analysis.

Detecting plagiarism in Arabic documents is particularly a challenging task because of the complex linguistic structure of Arabic language. In this paper, we are studying these three different models, in nature and required resources, for less-related languages such as English-Arabic.

Keywords: Cross-language plagiarism detection · CL-CNG method · CL-ASA method · T+MA method translation machine · Arabic-English plagiarism

1 Introduction

Plagiarism becomes one of the most important and serious forms of academic misconduct. It can be defined as the reuse of someone else's ideas, results, or words without acknowledging the original source [1].

Due to its size, the Web makes plagiarism detection a hard task, and it is even more difficult when is among documents written in different languages. According to Barròn-Cedeño et al. [7], Cross-Language Plagiarism Detection (CLPD) consists in

© Springer Nature Switzerland AG 2019
F. Khoukhi et al. (Eds.): AIT2S 2018, LNNS 66, pp. 207–213, 2019.
https://doi.org/10.1007/978-3-030-11914-0_22

discriminating semantically similar texts independent of the languages they are written in, when no reference to the original source is given [2]. CLPD case takes place when we deal with unacknowledged reuse of a text involving its translation from one language to another [2].

CLPD issue has acquired pronounced importance lately since semantic contents of a document can be easily and discreetly plagiarized through the use of translation (human or machine-based) [2]. Paraphrasing is a technique to modify the structure of an original sentence by changing the sentence's structure or replacing some of the original words with its synonym. Without proper citation or quotation marks, it is also considered as plagiarism [3].

Arabic language belongs to the Afro-Asian language group It has much specificity which makes it very different from other Indo-European languages. Detecting plagiarism in Arabic documents is particularly a challenging task and it becomes even harder, as translation is often a fuzzy process that is hard to search for, because of the complex linguistic structure of Arabic [4]. In spite of the fact that many researches were conducted on plagiarism detection in the last decades, those concerning the Arabic language text remain quite limited and addressed especially to monolingual plagiarism.

2   Related Work

Cross-language plagiarism detection confronts many difficulties especially when handling less common languages.

Potthast et al. [5] presents a multilingual retrieval model Cross-Language Explicit Similarity Analysis (CL-ESA) for the analysis of cross-language similarity. They did experiments on a multilingual comparable corpus (Wikipedia) and multilingual parallel corpus [1]. In [6] authors compare CL-ESA to other methods and they did find that CL-CNG achieves a better performance. Nevertheless, CL-Character N-grams will not be adequate for languages with unrelated syntax. Potthast et al. [6] define the steps of Cross-Language Plagiarism Detection process, put up some strategies of heuristic retrieval and evaluate the performance of the models for the detailed analysis.

The work by Barrón Cedeño et al. [7] relies on a statistical bilingual dictionary created from parallel corpora and on an algorithm for bilingual text alignment. The results of the experiments were satisfying and showed the similarity between the original documents and their plagiarized versions [1]. Multi-Lingual Plagiarism detector (MLPlag) [8] is a CLPD method based on the analysis of word positions. EuroWordNet is used to transform words into a language-independent representation. EuroWordNet may lead to difficulties during cross-language plagiarism detection, especially when handling less common languages. The work by Barrón Cedeño et al. [9] carry out the Kullback-Leibler distance to reduce the number of documents that must be compared against the suspicious document [1].

Although many studies were directed on plagiarism detection in the last years, those concerning the Arabic language text remain quite limited. Works in this area are those of Alzahrani et al. [10], Menai et al. [4] and others [11, 12]. All of them addressed the monolingual external approach.

AbdulJaleel et al. [13] works on statical transliteration for English-Arabic Cross Language Information retrieval (CLIR), authors worked with n-gram model and evaluate the statistically-trained model and a simpler hand-crafted model on a test set of named entities from the Arabic AFP-Corpus and demonstrate that they perform better than two online translation sources.

Bensalem et al. [14] Work on Arabic intrinsic plagiarism detection. They presented a set of preliminary experiments on intrinsic plagiarism detection in Arabic text using Stylysis tool and a small corpus. Their approach consists in testing whether some language-independent stylistic features are effective or not to discriminate between plagiarized and not plagiarized sentences, the results they found is that average word length and average sentence length are not reliable stylistic discriminator of Arabic text.

3 Cross-Language Character N-Gram

An n-gram is a subsequence of n items constructed from a given sequence. The idea can be traced to an experiment by Claude Shannon's work in information theory. His idea was that, from a given sequence of letters, it is possible to obtain the likelihood function of the appearance of the next letter. From a training corpus, it is easy to construct a probability distribution for the next letter with a history of size n. This modelization corresponds in fact to a Markov model of order n, where only the last n observations are used for prediction of the next letter.

CL-CNG model [15] is a model based on the syntax of documents, which uses character n-grams and offers remarkable performance for languages with syntactic similarities [16]. Their use is based on the simplifying assumption that given a sequence of k elements (k > n) the probability of the occurrence of an element in position i only depends on the previous n-1 elements.

Therefore

$$p(w_i|w_1, \ldots, w_{i-1}) = p(w_i|w_{i-n+1}, \ldots, w_{i-1}) \tag{1}$$

Similarity $sim(d_q, d')$ is estimated by the unigram language model [17]:

$$sim(d_q, d') = P(d'|d_q) = \prod_{q \in d_q} [\alpha P(q|d') + (1 - \alpha)P(q|C)] \tag{2}$$

Where dq is a suspicious document, $p(q|d')$ is the document level probability of term q in document d', α is a smoothing parameter and C denotes the entire collection [15].

The interest in using character n-grams comes from the possibilities they offer, particularly in the case of non-English languages [15, 18]. As it provides a substitution means to normalize word forms and it does not rely on language-specific processing, it can be applied to very different languages, even when linguistic information and resources are scarce or unavailable [19].

4   Cross-Language Alignment-Based Similarity Analysis

CL-ASA model, proposed by Barròn Cedeño et al. [7], to detect plagiarised text fragments (other models have been proposed for extraction of parallel sentences from comparable corpora). It is based on a statistical machine translation technology that combines probabilistic translation, using a statistical bilingual dictionary and similarity analysis [20]. In [16] CL-ASA model, an estimation of how likely is that d′ is a translation of dq is performed. It is based on the adaptation of the Bayes rule for MT:

$$p(d'|d_q) = \frac{p(d')p(d_q|d')}{p(d_q)} \tag{3}$$

$p(d_q)$ and d′ are independent, the probability is negligent. p(d′) is a language model probability; it describes the target language L′ in order to obtain grammatically acceptable translations.

The translation of a document (dq) from a language into another (L′), is not the concern of CL-ASA method, somewhat it focuses on retrieving texts written in L′ which are potential translations of dq. Multiple translations from a document (d) into L′ are possible, and it is uncommon to find a pair of translated texts d and d′ such that | d| = |d′| [20]. Consequently, Barròn Cedeño et al. [7] proposed to replace the language model (the one used in T+MA model) by the length model. This model depends on text's character lengths instead of language structures.

5   Translation Plus Monolingual Analysis

T+MA belongs to Machine translation-based systems followed by a monolingual similarity estimation; it simplifies the CLPD task by turning it into a monolingual problem. The first step of this approach is applying a language detector to determine the most likely language of dq. In the second step, the translation of all the documents into a common language, then monolingual analysis. T+MA includes web-based CL models and multiple translations [17]. Moreover, this method proved to be less sensitive than CL-ASA to the lack of resources. This due to the fact that it considers both directions of the translation model. Additionally, the language model, applied in order to compose syntactically correct translations, reduces the amount of wrong translations and, indirectly, includes more syntactic information in the process [20].

6   Comparison

This study generally aims to compare a wide variety of cross-language similarity models using several works in the same territory. The resources required by the models are integrally different:

T+MA requires the previous translation of all the texts, i.e. a complete MT system, which can be very costly for large collections. CL-ASA requires translation probabilities from aligned corpora, to estimate a bilingual dictionary and a length model, once the probabilities are estimated, cross-language similarity can be computed very fast, and CL-CNG is a simple model which does not depend on any resource, it requires only little language specific adjustments, e.g. alphabet normalization by removal of diacritics [17, 20].

CL-CNG offers remarkable performance for languages with syntactic similarities [16], particularly in the case of non-English languages.

Since it provides a surrogate means to normalize word forms and it does not rely on language-specific processing, it can be applied to multilingual documents without translation, the best results were achieved for the languages sharing similar syntactic structure and international lexicon (e.g., related European language pairs) [2], even when linguistic information and resources are scarce or unavailable.

Nevertheless, when prolonging its use to Cross-Language Plagiarism Detection for Arabic language the CL-CNG is not an efficient approach to corpus-based Arabic word conflation [21], an extra translation phase is needed [19].

CL-ASA, as opposed to CL-CNG, can be applied for distant language pairs with alphabet and syntax unrelated, as pointed out in Barròn Cedeño (2012). It provides high precision values; particularly in case of long texts, short plagiarism cases appear to be the hardest to detect. Results obtained using CL-CNG, despite of not being influenced by the length and nature of plagiarism, turned out to be the worst ones [2]. On the basis of the experiments performed authors conclude that T+MA and CL-CNG can be considered as recall-oriented systems and CL-ASA as a precision-oriented one (Barròn Cedeño [17]). CL-ASA obtains better results in exact (human) translations.

However, most of the language pairs used in throws works are related, because a large proportion of their vocabularies share common roots and they have common predecessors. The closeness among languages is an important factor [20]; In fact, the lack syntactic relation between the English-Arabic pair may cause a performance degradation for CL-CNG, and for CL-ASA to a lesser extent. As opposed to CL-ASA, T+MA work better with longer documents (Due to the use of length model) [2].

Although T+MA is a simple approach that reduces the cross-language similarity estimation to a translation followed by a monolingual process, and performed better than the other models because it does not depend neither on corpora nor on syntactic/lexical similarities between languages [2]. It proved to be less sensitive than CL-ASA to the lack of resources. This due to the fact that it considers both directions of the translation.

Besides, the language model, applied to compose syntactically correct translations, reduces the number of wrong translations and integrate more syntactic information in the process. On the contrary, CL-ASA only considers one direction translation model and completely disregards syntactic relations between the texts [20]. However, T+MA is a computationally expensive method and there is still a lack of good automatic translators for most language pairs, in addition, mistakes of translations during the language normalisation stage may cause lower precision levels of the model [2, 20].

CL-ASA is a competitive model, even using limited dictionaries, it works much better on "exact" translations than on comparable documents and it is used on language pairs whose alphabet or syntax are unrelated like our case but translated plagiarism cases must be exact copies of the original source [22]. CL-ASA is roughly comparable to T+MA, without relying on a translation module, but with much higher precision, causing the workload of the human reviewer to decrease [5]. CL-CNG model can be used to detect plagiarism between the languages with similar syntactic, it is used in many research for monolingual plagiarism detection in Arabic [11] and gives remarkable results, but it is not appropriate for less-related languages in general and especially for Arabic-English case. Whereas T+MA performs better than the previously proposed models [20].

7 Conclusion

Automatic plagiarism detection models aim to find potential cases of unauthorised text reuse. In this paper, we studied three models of CLPD for English-Arabic Languages: cross-language alignment-based similarity analysis, cross-language character n-grams, and translation plus monolingual analysis. CL-CNG can be directly operationalized, necessitates only little language specific adjustments and offers remarkable performance for languages with syntactic similarities, but CL-CNG shows not be appropriate for the kind of language pairs we work with (i.e. English-Arabic). CL-ASA requires translation probabilities from aligned corpora, this model performs very well with professional and automatic translations, and it can be used on language pairs with unrelated syntax. However, CL-ASA is sensitive to the amount of information it can exploit and it performance tend to decrease for long documents. Although T+MA performs better than the previously proposed models. It requires the previous translation of all the texts, this method attested to be less sensitive the lack of resources and work better with longer documents comparing it with the CL-ASA.

As future work, we will try to improve or adjust one of the models to be suitable for English-Arabic and French-Arabic, and tested on a specific corpus that will be constructed taking on consideration all Arabic language specificities in CLPD.

References

1. Pereira, R.C., et al.: A new approach for cross-language plagiarism analysis. In: Multilingual and Multimodal Information Access Evaluation, pp. 15–26. Springer, Heidelberg (2010)
2. Danilova, V.: Cross-language plagiarism detection methods. In: RANLP, pp. 51–57 (2013)
3. Kent, C.K., Salim, N.: Web based cross language semantic plagiarism detection. In: 2011 IEEE Ninth International Conference on Dependable, Autonomic and Secure Computing (DASC), pp. 1096–1102. IEEE, December 2011
4. Menai, M.E.B.: Detection of plagiarism in Arabic documents. Int. J. Inf. Technol. Comput. Sci. (IJITCS) 4(10), 80 (2012)
5. Potthast, M.: Wikipedia in the pocket: indexing technology for near-duplicate detection and high similarity search. In: Proceedings of the 30th Annual International ACM SIGIR Conference on Research and Development in Information Retrieval, pp. 909–909. ACM, July 2007
6. Potthast, M., Barrón-Cedeño, A., Stein, B., Rosso, P.: Cross-language plagiarism detection. Lang. Resour. Eval. 45(1), 45–62 (2011)
7. Barrón-Cedeno, A., et al.: On cross-lingual plagiarism analysis using a statistical model. In: PAN (2008)

8. Ceska, Z., Toman, M., Jezek, K.: Multilingual plagiarism detection. In: Artificial Intelligence: Methodology, Systems, and Applications, pp. 83–92. Springer, Heidelberg (2008)

9. Barrón-Cedeno, A., et al.: Reducing the plagiarism detection search space on the basis of the Kullback-Leibler distance. In: Computational Linguistics and Intelligent Text Processing, pp. 523–534. Springer, Heidelberg (2009)

10. Alzahrani, S., Salim, N.: Fuzzy semantic-based string similarity for extrinsic plagiarism detection. Braschler Harman **1176**, 1–8 (2010)

11. Zitouni, A., et al.: Corpus-based Arabic stemming using N-grams. In: Information Retrieval Technology, pp. 280–289. Springer, Heidelberg (2010)

12. Siddiqui, M.A., et al.: Developing an Arabic plagiarism detection corpus. Grant No. 11-INF-1520-03

13. AbdulJaleel, N., Larkey, L.S.: Statistical transliteration for English-Arabic cross language information retrieval. In: Proceedings of the Twelfth International Conference on Information and Knowledge Management. ACM (2003)

14. Bensalem, I., Rosso, P., Chikhi, S.: Intrinsic plagiarism detection in Arabic text: preliminary experiments. In: II Spanish Conference on Information Retrieval (CERI 2012) (2012)

15. Mcnamee, P., Mayfield, J.: Character n-gram tokenization for European language text retrieval. Inf. Retr. **7**(1–2), 73–97 (2004)

16. Franco-Salvador, M., Gupta, P., Rosso, P.: Cross-language plagiarism detection using a multilingual semantic network. In: Advances in Information Retrieval, pp. 710–713. Springer, Heidelberg (2013)

17. Barrón-Cedeño, A., Gupta, P., Rosso, P.: Methods for cross-language plagiarism detection. Knowl. Based Syst. **50**, 211–217 (2013)

18. McNamee, P., Mayfield, J.: JHU/APL experiments in tokenization and non-word translation. In: Comparative Evaluation of Multilingual Information Access Systems, pp. 85–97. Springer, Heidelberg (2004)

19. Vilares, J., Oakes, M.P., Vilares, M.: Character N-grams translation in cross-language information retrieval. In: Natural Language Processing and Information Systems, pp. 217–228. Springer, Heidelberg (2007)

20. Barrón-Cedeno, A., Rosso, P., Agirre, E., Labaka, G.: Plagiarism detection across distant language pairs. In: Proceedings of the 23rd International Conference on Computational Linguistics, pp. 37–45. Association for Computational Linguistics, August 2010

21. Khreisat, L.: Arabic text classification using N-gram frequency statistics a comparative study. In: Conference on Data Mining, DMIN 2006 (2006)

22. Franco-Salvador, M.: Cross-language plagiarism detection using a multilingual semantic network (2013)

Stock Price Forecasting: Improved Long Short Term Memory Model for Uptrend Detecting

Yassine Touzani[✉], Khadja Douzi, and Fadoul Khoukhi

Computer Laboratory of Mohammedia (LIM),
Faculty of Sciences and Technology Mohammedia (FSTM),
Hassan II University of Casablanca, 20650 Mohammedia, Morocco
touzayassine@gmail.com, kdouzi@yahoo.fr, khoukhif@gmail.com

Abstract. Stock price forecasting is one of the most challenging activities for traders and financial analysts due to the high volatility of stock market data. Investing in the stock market is often associated with a significant risk, hence the need for a forecasting model to minimize it. To maximize profit, investors must anticipate upward trends to place sales orders at the highest possible price. This article aims to present a new model for uptrend detecting for a given horizon. Our proposed model consists of successive phases: the first phase is for selecting stock market that have returns normally distributed. Note that the returns in our case reflect the evolution of prices average in a given period compared to the current price. Then a filtration that aims to keep only the stock price that have the average ratio over standard deviation as high as possible. Subsequently, we create our portfolio using selected stock market. In the next phase, we will build classes by comparing returns to a given limit. The next step is for training and testing two classifier: LSTM classifier and nearest neighbor classifier. The last phase is to cross the results provided by the two models and decide on the value to be predicted according to a decision rule. Experience shows that our proposed model gives very promising results and its accuracy to predict uptrend is high.

Keywords: Stock market prediction · Deep learning · LSTM ·
Nearsest neighbor · Classification · Machine learning ·
Normal distribution

1 Introduction

Accoridng to market hypothesis stock price behavior is random and its prediction is impossible [3] stock price forecasting is a very challenging activity for the searcher in this field, because of its importance and its difficulty due to its dependance on several factors (financial, poltical, emotional etc). In this paper we present a new hybrid model for uptrend anticipating. From all stock market we will keep only those who have a return normally distributed. After we introduce

© Springer Nature Switzerland AG 2019
F. Khoukhi et al. (Eds.): AIT2S 2018, LNNS 66, pp. 214–219, 2019.
https://doi.org/10.1007/978-3-030-11914-0_23

a new filter based on mean and standard deviation to exclude stock market that have a return with a high standard deviation comparing to its average. Thereafter we create a portfolio, then we define the threshold from which we will consider an uptrend and we label the classes accordingly. Finally data transformation is performed using the min-max normalization [5] and long short term memory model and nearest neighbor classifier are trained and tested. The prediction obtained from the two models are crossed, a decision rule is defined to predict final output. The model gives good results and its ability to identify future price increases is acceptable. The rest of this paper is organized as follows: The related work will be presented in Sect. 2, followed by our proposed model description in Sect. 3. Section 4 will focus on the experiments and results obtained, finally the conclusion is drawn in Sect. 5.

2   Related Work

One of the most popular technique in time series analysis and detecting trend and seasonality is Autoregressive integrated moving average ARIMA [1]. Currently deep learning techniques are increasingly used for the analysis of financial time series [4]. Long short term memory LSTM are among the most used neuron networks in this field. In [6] Vargas et al. use LSTM with technical indicators to forecast SP500 index. To forecast stock price volatility. Kim and Won present in [7] a hybrid model based on LSTM network and several GARCH model. in [8] Abe and Nakayama are implementing deep learning techniques to predict returns in the Japanese stock market for horizon one month ahead and find that deep neural network outperform the shallow neural network in stock market forecasting. In order to predict the daily closing price of the Shanghai Composite Index, Yan and Ouyang present in [9] a model that combines wavelet analysis and LSTM. The wavelet analysis is used to perform multiple decompositions of the original time series into low frequency and high frequency signals. Low frequencies reflect overall trend and high frequencies consider as noisy data that will be ignored. They obtained an approximate signal of the original financial time series. Then LSTM neural network is applied for prediction task.

3   Proposed Model Description

The proposed model is composed of several steps that we will detail in this section.

3.1   Data Preparation

To mitigate stock price volatility we will analyze return instead of initial stock market price. The return formula is given by:

$$r(t) = \frac{y(t+h) - y(t)}{y(t)} \tag{1}$$

where y(t) is closing price at the moment t. It shows the evolution of the price between time t and t + h relative to the reference price y(t). For our model, we will be interested in the evolution of the price between the closing price y(t) and the average price of the next h days. The return formula will therefore be slightly different and will be given by:

$$r(t) = \frac{m(t) - y(t)}{y(t)} \tag{2}$$

where m(t) is moving average from (t+1) to t+h given by formula:

$$m(t) = \frac{1}{h} * \sum_{i=1}^{h} y(t + i) \tag{3}$$

h is a given horizon.

3.2   Stock Market Selection

Working with data whose distribution is known is more convenient than data from an unknown distribution, We will test for each stock market the normality of the returns by using the Kolmogorov-Smirnov test with a high significance level to accept null hypothesis (H0: Data is drawn from normal distribution). with a high acceptance level we accept the risk of rejecting share whose return are normally distributed (Error type I is high) but we will be confident that the selected share have returns normally distributed (Error type II is low). After selecting stocks, we will eliminate stocks with low profit or a fairly high risk. The average return is the expectation of profit it need to be relatively high, the standard deviation meanwhile can quantify the risk of the stock market it must be relatively small. A natural filter is introduced given by

$$filtr = \frac{\mu}{\sigma} \tag{4}$$

where μ and σ are the return average and standard deviation respectively. Only stock score above a given threshold will be kept.

3.3   Class Definition

In this step we form a portfolio based on the selected shares, we give to each stock market a weight according to its score. Then we define a threshold, if the portfolio returns are greater than this limit the class is labeled up (a potential uptrend in the next h days) otherwise the class will be labeled as neutral.

3.4   Classifier Implementation

At this last phase we will build the classification features, normalize the data and train two classifier LSTM classifier and nearest neighbor classifier. At the end we will cross the results obtained, the prediction of a class 1 will only be valid if both models could identify it otherwise the predicted class will be 0.

4   Experiment and Results

4.1   Dataset

Our data extracted from Yahoo finance web and correspond to the trading information of SP&500 index going from 2009 until 2018. Data from 2009 to 2016 will be used as training data (including validating dataset) and data up to 2016 will be consider as test data. Forecasting horizon is set to 20 days. Stock market return is calculated according to the given formula in Eq. 2, then normality test with significant level (set to 0.15) is performed. Figure 1 show an example of stock that verify normality test and Table 2 summarize the output results of stock market selection step. We set the threshold for the filtr value to 0.2, all stock with a filtr value less than 0.2 will be discarded. Then The Portfolio is built, after binary classes are defined by comparing return to 5% (given threshold to identify potential uptrend). In order to train the two classifier LSTM and nearest neighbor, features are built from historical trading data. Once the features are finalized we train:

- LSTM model (initial training data partitioned into new trainnig and validating data). Validating data will used to optimize model paramters and avoid the underfitting and overfitting. Figure 2 proves this, we find that loss functions decreases for both training and validating data.
- Nearest neighbor model using minkowski distance.

Finally we cross the results provided from both models and we decide to predict class 1 if both model predict 1 otherwise 0 will be predicted.

4.2   Metrics

Precision and recall indicators will be used to evaluate the model performance for a specific class (0 or 1) the Table 1 summarize all possibilities for one prediction. Precision and recall formula is given by [2]

$$precision = \frac{TP}{TP + FP} \quad recall = \frac{TP}{TP + FN} \tag{5}$$

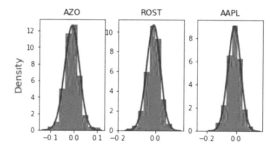

Fig. 1. Return distribution

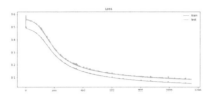

Fig. 2. LSTM Loss function

Precision indicator measure the reliability of the model to predict a class. (for a given class c it indicate the ratio of good predicted class c over the all predicted class c) while recall indicator give information about the ability of model to not skip a given class.

4.3 Proposed Model Results and Discussion

Table 3 show the confusion matrix of LSTM model and Table 4 present the onfusion matrix after improvement. From these tables we can clearly see the improvement. The LSTM model alone predict a wrong uptrend 87 times which is problematic (precision = 61%) by including the NN improvement the number of wrong uptrend predicted decrease to 35. Even the number of correct predicted uptrend class decrease from 137 to 87 after improvement, the value of the precision indicator is improved to 71.3%.

Table 1. Prediction categories

Predict data	Real data	
	Class = c	Class != c
Class = c	TP	FP
Class != c	FN	TN

Table 2. Stock market selection results top 3 records

Stock	Mean	Std	Measure
AZO	0.010261	0.02611	0.392975
ROST	0.011397	0.037215	0.306257
AAPL	0.012619	0.043979	0.286927

Table 3. LSTM model confusion matrix

Real data	Predicted data	
	Class = 0	Class = 1
Class = 0	402	83
Class = 1	0	137

Table 4. Improved LSTM model confusion matrix

Real data	Predicted data	
	Class = 0	Class = 1
Class = 0	450	35
Class = 1	50	87

5   Conclusion

In this article we have proposed a hybrid model based on the lstm and NN class-fier (applied to stock market that have return normally distributed) to identify only future uptrend. Future work will be dedicated to downtrends identification to complete the presented paper.

References

1. Box, G.E.P., Jenkins, G.M., Reinsel, G.C.: Time Series Analysis Forecasting and Control. Wiley, New York (2008)
2. Olson, D.L., Delen, D.: Advanced Data Mining Techniques. Springer, Heidelberg (2008)
3. Patel, J., Shah, S., Thakkar, P., Kotecha, K.: Predicting stock market index using fusion of machine learning techniques. Expert Syst. Appl. **42**, 2162–2172 (2015)
4. LeCun, Y., Bengio, Y., Hinton, G.: Deep learning. Nature **521**, 436–444 (2015)
5. Bao, W., Yue, J., Rao, Y.: A deep learning framework for financial time series using stacked autoencoders and long-short term memory. PLOS ONE **12** (2017). https://doi.org/10.1371/journal.pone.0180944
6. Vargas, M.R., de Lima, B.S., Evsukoff, A.G.: Deep learning for stock market prediction from financial news articles. In: 2017 IEEE International Conference on Computational Intelligence and Virtual Environments for Measurement Systems and Applications (CIVEMSA), pp. 60–65. IEEE (2017)
7. Kim, H.Y., Won, C.H.: Forecasting the volatility of stock price index: a hybrid model integrating LSTM with multiple GARCH-type models. Expert Syst. Appl. **103**, 25–37 (2018)
8. Abe, M., Nakayama, H.: Deep learning for forecasting stock returns in the cross-section. In: Phung, D., Tseng, V.S., Webb, G.I., Ho, B., Ganji, M., Rashidi, L. (eds.) Advances in Knowledge Discovery and Data Mining, pp. 273–284. Springer International Publishing, Cham (2018)
9. Yan, H., Ouyang, H.: Financial time series prediction based on deep learning. Wireless Pers. Commun. **102**, 683–700 (2018)

Using Genetic Algorithm to Improve Classification of Imbalanced Datasets for Credit Card Fraud Detection

Ibtissam Benchaji[✉], Samira Douzi, and Bouabid El Ouahidi

Department of Computer IPSS, Faculty of Sciences,
Mohammed V University, Rabat, Morocco
b.ibtissam@gmail.com, samiradouzi8@gmail.com,
Bouabid.ouahidi@gmail.com

Abstract. With the growing usage of credit card transactions, financial fraud crimes have also been drastically increased leading to the loss of huge amounts in the finance industry. Having an efficient fraud detection method has become a necessity for all banks in order to minimize such losses. In fact, credit card fraud detection system involves a major challenge: the credit card fraud data sets are highly imbalanced since the number of fraudulent transactions is much smaller than the legitimate ones. Thus, many of traditional classifiers often fail to detect minority class objects for these skewed data sets. This paper aims first: to enhance classified performance of the minority of credit card fraud instances in the imbalanced data set, for that we propose a sampling method based on the K-means clustering and the genetic algorithm. We used K-means algorithm to cluster and group the minority kind of sample, and in each cluster we use the genetic algorithm to gain the new samples and construct an accurate fraud detection classifier.

Keywords: Fraud detection · Imbalanced dataset · K-means clustering · Genetic · Programming · Autoencoder

1 Introduction

Credit card fraud is a widely increasing problem in the credit card industry, particularly in the online sector. These illegal activities that aim to obtain goods without paying, or to gain illegitimate funds from an account, have caused severe damage to the users and the service provider.

Moreover, the main challenge of credit card fraud problem is to detect frauds in a huge dataset where the legal transactions are more and the fraudulent transactions are minimum or close to negligible. Hence, Developing Fraud Detection System (FDS) based on imbalanced datasets is one of the major challenges in machine learning.

A dataset is called imbalanced when the number of negative (majority) instances outnumbers the amount of positive (minority) class instances. Such imbalanced sets require additional precautions because the prediction accuracy of standard machine learning techniques, especially for the minority class which is the class of interest, tends to be lower. In real world applications, classification accuracy of the smaller class

© Springer Nature Switzerland AG 2019
F. Khoukhi et al. (Eds.): AIT2S 2018, LNNS 66, pp. 220–229, 2019.
https://doi.org/10.1007/978-3-030-11914-0_24

is critically important because the minority class usually is the class of great interest such as cases of fraud. Thus, misclassifying the minority class has much higher cost compared to misclassifying a majority class instance. Notice that, when the system predicts a transaction as fraudulent when in fact it is not (false positive), the financial institution has an administrative cost, as well as a decrease in customer satisfaction. On the contrary, when the system does not detect a fraudulent transaction (false negative), the amount of that transaction is lost.

In this work, we propose the generation method of imbalanced data set's minority class, by using K-means clustering and genetic algorithm, new samples can be obtained through crossover on the basis of cluster as the complement of the minority class samples.

The remainder of this paper is organized as follows. Section 2 introduces the imbalanced data sets challenge faced by detection systems. Section 3 presents some required concepts and gives all the details of our proposed approach. Finally, Sect. 4 presents conclusions and suggestions for future work.

2   Classification of Imbalanced Datasets

First, as introduced in Sect. 1, the major challenge to be addressed when designing a FDS is handling the class imbalance, since legitimate transactions far outnumber the fraudulent ones. In fact, Class distribution is extremely unbalanced in credit card transactions, since frauds are typically less than 1% of the overall transactions, as shown in [1].

In this regard, various approaches have been proposed to deal with imbalanced datasets issue and improve the performance of predictive modeling. These approaches could be mainly divided into two categories. The first category of methods solve the problem at data level, that is, data resample techniques are performed to directly alter the original dataset which is not balanced by removing samples from the majority class (undersampling) or replicating training samples of the minority class (oversampling) [2]. Advanced oversampling methods like SMOTE [3] generate synthetic training instances from the minority class by interpolation, instead of sample replication. However, these techniques present significant drawbacks, such as undersampling may lose some potential information, and oversampling may lead the overfitting.

The second main type of approaches that considers classification of imbalanced data sets at algorithm level is known as ensemble learning. Ensemble methods include bagging and boosting strategies that aim to combine multiple classifiers in order to reduce the data variance. The ideas of Bagging (Bootstrap Aggregating) [4] is to split the majority class into multiple sub sets, and for each sub set, training a basic weak classifier together with the minority class, then integrate these classifiers into a strong classifier. Bagging is proposed to use random sampling technique to generate empirical distribution so as to approximate the real distribution of the data set. The AdaBoost (Adaptive Boosting) [5] talks about the misclassified samples that generated by the former basic classifier will be augmented by assigning weights. The weighted data sets will be send to next basic classifier. AdaBoost (Adaptive Boosting) focus on the misclassified samples that generated by the former basic classifier to be augmented by assigning weights. The weighted data sets will be send to next basic classifier to reduce the total bias error. The weighting strategy of AdaBoost is equivalent to resampling the data space [6], which are applicable to most classification systems without changing their learning methods. Besides, it could eliminate the extra learning cost for exploring the optimal class distribution and representative samples [7]. Moreover, compared with the method of eliminating samples from data set, it reduces the information loss, overfitting risk and bias error of a certain classification learning method [8].

3   Proposed Methodology

In an effort to improve the performance of imbalanced data classification when designing Fraud Detection System, a new oversampling strategy using K-means cluster and genetic algorithms to create synthetic instances from the minority class is proposed in this paper. It is designed to overcome the limitations of existing sampling methods such as information loss or overfitting. Figure 1 shows the big picture of the proposed method. Our suggested approach generates new minority class instances in each cluster and merges them with the original dataset to gain new training sets in order to construct an accurate fraud detection classifier.

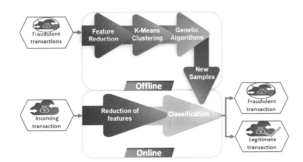

Fig. 1. The illustration of credit card fraud detection system

3.1   Financial Fraud Data and Feature Selection Using Autoencoder

Creating domain expertise features has proven to be an integral part of building predictive models in detecting card fraud; banks collect a large amount of information about card accounts, account holders, and transactions. However, not all of this data directly specifies important predictors such as consumer spending habits throughout time, or variance in predictors, etc.

The basis of credit card fraud detection lies in the analysis of cardholder's spending behavior. This spending profile is analyzed using optimal selection of variables that capture the unique behavior of a credit card and detect very dissimilar transactions within the purchases of a customer. Also, since the profile of both a legitimate and fraudulent transaction tends to be constantly changing, optimal selection of variables that greatly differentiates both profiles is needed to achieve efficient classification of credit card transaction.

Therefore Feature selection is a fundamental preprocessing step in fraud detection systems, to select the optimal subset of relevant features by removing redundant, noisy, and irrelevant features from the original dataset, and decrease the computational cost without a negative effect on the classification accuracy. Thus, Autoencoder has been seen as one of the best technique used for feature selection in many applications such as image analysis. For example, it has been utilized in medical imaging to reconstruct training samples in order to create compressed sensing images from MRI and CT [9].

Autoencoder can be regarded as a special form of neural network designed for unsupervised learning [10, 11]. The aim of an Autoencoder is to transform inputs into outputs with the least possible amount of deviation [12].

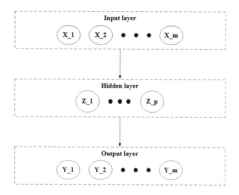

Fig. 2. The Autoencoder architecture

The basic version of an Autoencoder consists of three fully connected layers, i.e., one input layer, one hidden layer and one output layer. The reconstruction ability can be improved by introducing more hidden layers and hidden neurons. As shown in Fig. 2, the input (i.e., denoted as X) and output (i.e., denoted as Y) are set identical with a dimension of n and m, where n is the number of observations and m is the variable number. An Autoencoder consists of an encoder and a decoder. The encoder transforms the input data into high-level features (i.e., denoted as Z), while the decoder tries to reconstruct the input data using high-level features.

The Autoencoder is applied on the set of features that contains credit card transactions to generate the robust and discriminative features for fraudulent instances.

3.2   K-Means Clustering Method

Clustering is a process of partitioning a set of samples into a number of groups without any supervised training. The goal of this technique is to categorize the records of a dataset in such a way that similar records are grouped together in a cluster and dissimilar records are placed in different clusters. The more the similarity among the data in clusters, more the chances of particular data-items to belong to particular group. Thus, K-Means is a heuristic clustering algorithm based on distance measure that partitions a data set into K clusters by minimizing the sum of squared distance in each cluster.

The algorithm consists of three main steps: (i) initialization by setting center points (or initial centroids) with a given K, (ii) Dividing all data points into K clusters based on K current centroids, and (iii) updating K centroids based on newly formed clusters. K-Means algorithm converges after several iterations of repeating steps (ii) and (iii). These steps are shown as Fig. 3.

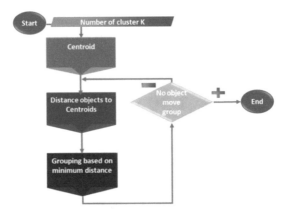

Fig. 3. K-means clustering flow chart

3.2.1   The K-Means Clustering Algorithm Is Specified as Follows

- Let X = {x1, x2, x3,…….., xn} be the set of data points and V = {v1, v2,……., vc} be the set of centers.
- Step 1: Select 'c' cluster centers randomly.
- Step 2: Calculate the distance between each data point and cluster centers using the Euclidean distance metric as follows

$$Dist_{XY} = \sqrt{\sum_{k=1}^{m} (X_{ik} - X_{jk})^2}$$

- Step 3: Data point is assigned to the cluster center whose distance from the cluster center is minimum of all the cluster centers.
- Step 4: New cluster center is calculated using:

$$V_i = \left(\frac{1}{Ci}\right) \sum_{1}^{Ci} xi$$

Where, 'ci' denotes the number of data points in this cluster.
- Step 5: The distance between each data point and new obtained cluster centers is recalculated.
- Step 6: If no data point was reassigned then stop, otherwise repeat steps from 3 to 5.

In the proposed paper, the k-means algorithm is used to split minority class of fraud instances into clusters according to their similarities and generate new samples in these clusters. Thus, new samples are created from samples that are as much similar as possible. Thus, every cluster can obtain certain proportion of new samples, which can guarantee that new samples of whole minority class have better coverage and representation. Figure 4 presents the schematic diagram of this step.

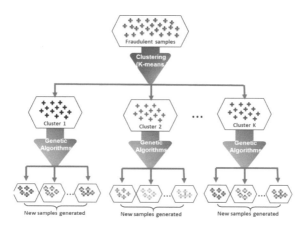

Fig. 4. GA generation process

3.3   Genetic Algorithms

Genetic algorithms were inspired from Darwin's theory of evolution and were pioneered by John Holland [13]. A genetic algorithm can be defined as a search algorithm based on the mechanics of natural selection and natural genetics [14]. Genetic algorithms have at least the following elements in common: Populations of chromosomes, selection according to fitness, crossover to produce offspring, and random mutation of a new offspring [15]. In a broader usage of the term, a genetic algorithm is any population-based model that uses selection and recombination operators to generate new sample points in a search space. Figure 5 depicts a schematic view of different steps involved.

In further details, the algorithm starts with a population of fraudulent transaction samples. Each sample within the population is coded in a so-called chromosome into a specific type of representation (i.e. binary, decimal, float, etc).

Each chromosome (sequences of genes) is assigned a "fitness" according to how valid new individuals' property features are, based on a given fitness function. Fitness is calculated in the evaluation step.

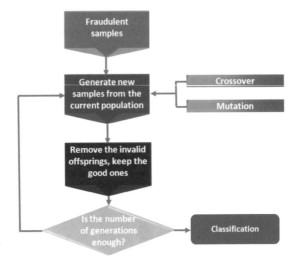

Fig. 5. GA flow

While the termination condition of number of generations is not met, the processes of selection, recombination, mutations and fitness calculations are done. Selection process chooses individuals from population for the process of crossover. Recombination (or crossover) is done by exchanging a part (or some parts) between the chosen individuals, which is dependent on the type of crossover (Single point, Two points, Uniform, etc). Mutation is done after that by replacing few points among randomly chosen individuals. Then fitness has to be recalculated to be the basis for the next cycle.

3.3.1 Pseudo Code of Genetic Algorithm

- Initialize the population
- Evaluate initial population
- Repeat
    Perform competitive selection
    Apply genetic operators to generate new solutions
    Evaluate solutions in the population
- Until some convergence criteria is satisfied

3.3.2 Selection Process
The selection process is used for choosing the best individuals from the population which lead to the further breeding in the next generation. The selection operation takes the current population and produces a 'mating pool' which contains the individuals which are going to reproduce. The higher of the fitness value, the higher the probability to be selected and copied many times. There are three common types of selection operator, Fitness proportional/Roulette Wheel Selection [16], Stochastic Universal Sampling [17] and Binary Tournament Selection Tournament [16].

3.3.3   Crossover Process

Crossover is also a genetic operator which is succeeded by the selection. It takes two individuals and cuts their chromosome strings at some chosen position to produce two "head" segments and two "tail" segments. The tail segments are then swapped over to produce two new full length chromosomes as depicted in Fig. 6. Each of the two offspring inherits some genes from each parent. Crossover is made with the hope that new chromosomes will contain good parts of old chromosomes. As a result, the new chromosomes are expected to be better. If crossover is performed, the genes between the parents are swapped, and the offspring is made from parts of both parents' chromosomes. If no crossover is performed, the offspring is an exact copy of its parents.

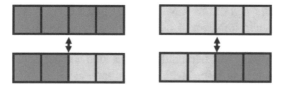

Fig. 6. Crossover

3.3.4   Mutation Process

The last process in genetic algorithm is the mutation process. It is mainly used to maintain heterogeneity in the population. It alters some of the genes to create a unique and more fit offsprings in the population. Mutation is applied to each child individually after the crossover that alters each gene with a low probability, typically in the range 0.001 and 0.01, and modifies elements in the chromosomes [16]. Mutation is often seen as providing a guarantee that the probability of searching any given string will never be zero, it acts as a safety net to recover the good genetic material that may be lost through the action of selection and crossover. Mutation prevents the GA from falling into local extremes and provides a small amount of random search that helps ensure that no point in the search space has a zero probability of being examined. If mutation is performed, one or more parts of a chromosome are changed, and if there is no mutation, the offspring is generated immediately after the crossover (or directly copied) without any change [16] (Fig. 7).

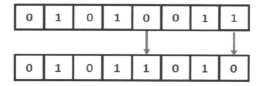

Fig. 7. Mutation

In the proposed paper, a number of genetic operations are applied on group to code the samples of each cluster into chromosome, then crossover operator is used to get a new generation group. New individuals will inherit the individual's characteristic of father's generation, but not simple duplication. It aims to get a new balanced training set that allow classifiers to be more accurate in detecting fraudulent transactions.

4   Conclusion

In this study, a new method for data generation of imbalanced data set's minority class was proposed to enhance fraud detection in e-banking by using K-Means clustering and genetic algorithm as an oversampling strategy.

Although genetic algorithms have been applied in many areas, our application domain aims to handle imbalanced data set issue by generating new minority class instances to gain new training sets. Applying this algorithm into bank credit card fraud detection system aims to reduce fraudulent transaction and decrease the number of false alert. A further work is to implement this approach using python programming language, this will allow us to validate our work and produce pertinent experimental results.

References

1. Dal Pozzolo, A., Johnson, R.A., Caelen, O., Waterschoot, S., Chawla, N.V., Bontempi, G.: Using HDDT to avoid instances propagation in unbalanced and evolving data streams. In: Proceedings of the International Joint Conference on Neural Networks, pp. 588–594 (2014)
2. Dal Pozzolo, A., Caelen, O., Bontempi, G.: When is undersampling effective in unbalanced classification tasks? In: Machine Learning and Knowledge Discovery in Databases. Springer, Cambridge (2015)
3. Chawla, N., Bowyer, K., Hall, L.O., Kegelmeyer, W.P.: SMOTE: synthetic minority over-sampling technique. Artif. Intell. Res. **16**, 321–357 (2002)
4. Breiman, L.: Bagging predictors. Mach. Learn. **24**(2), 123–140 (1996)
5. Yoav, F., Schapire, R.E.: Experiments with a new boosting algorithm. In: Machine Learning: Proceedings of the Thirteenth International Conference, pp. 148–156 (1996)
6. Sun, Y., Kamel, M.S., Wong, A.K., Wang, Y.: Cost-sensitive boosting for classification of imbalanced data. Pattern Recogn. **40**(12), 3358–3378 (2007)
7. Ali, A., Shamsuddin, S.M., Ralescu, A.L.: Classification with class imbalance problem: a review. Int. J. Adv. Soft. Comput. Appl. **7**(3), 176–204 (2015)
8. He, H., Garcia, E.A.: Learning from imbalanced data. IEEE Trans. Knowl. Data Eng. **21**(9), 1263–1284 (2009)
9. Mehta, J., Majumdar, A.: RODEO: robust de-aliasing autoencoder for real-time medical image reconstruction. Pattern Recogn. **63**, 449–510 (2017)
10. Zhuang, F., et al.: Representation learning via semi-supervised autoencoder for multi-task learning. In: EEE International Conference on Data Mining (2015)
11. úlrsoy, O., Alpaydõn, E.: Unsupervised feature extraction with autoencoder trees. Neurocomputing (2017). https://doi.org/10.1016/j.neucom.2017.02.075

12. Douzi, S., Amar, M., El Ouahidi, B.: Advanced phishing filter using autoencoder and denoising autoencoder. In: Proceedings of the International Conference on Big Data and Internet of Thing, pp. 125–129 (2017)
13. úlrsoy, O., Holland, J.: Adaptation in Natural and Artificial Systems. University of Michigan Press, Ann Arbor (1975)
14. Goldberg, D.: Computer-aided gas pipeline operation using genetic algorithms and rule learning. Ph.D. thesis. University of Michigan, Ann Arbor (1983)
15. Ben Amor, H., Rettinger, A.: Intelligent exploration for genetic algorithms: using self-organizing maps in evolutionary computation. In: Proceedings of the 2005 Conference on Genetic and Evolutionary Computation, pp. 1531–1538 (2005)
16. Goldberg, D.E.: Genetic Algorithms in Search, Optimization and Machine Learning, 1st edn. Addison-Wesley Longman Publishing Co., Inc., Boston (1989)
17. Baker, J.E.: Reducing bias and inefficiency in the selection algorithm. In: Proceedings of the Second International Conference of Genetic Algorithms and Their Application, pp. 14–21 (1987)

Intelligent Vehicle Routing System Using VANET Strategy Combined with a Distributed Ant Colony Optimization

Elgarej Mouhcine$^{(\boxtimes)}$, Khalifa Mansouri, and Youssfi Mohamed

Laboratory SSDIA, ENSET Mohammedia University Hassan II,
Mohammedia, Morocco
`mouhcine.elgarej@gmail.com`, `khamansouri@hotmail.com`, `med@youssfi.net`

Abstract. This work shows the impact of traffic road congestion which in volves driver frustration, air pollution and costs important money in fuel consumption by vehicles. Proposing an efficient strategy to reduce traffic congestion is an important challenge in which we need to take into consideration the unpredictable and the dynamic infrastructure of the road network. With the advances in computing technologies and communications protocols we can fetch any type of data from several entities in realtime about the traffic road congestion on each road based on: electronic toll collection system (ETCS), vehicle traffic routing system (VTRS), intelligent transportation system (ITS) and traffic light signals (TLS). This study introduces a new distributed strategy that aims to optimize traffic road congestion in realtime based on the Vehicular adhoc network (VANET) communication system and the techniques of the Ant colony optimization (ACO). The VANET is used as a communication technology that will help us to create a channel of communication between several vehicles. In the other hand, the techniques of the ACO is used to compute the shortest path that can be followed by the driver to avoid the congested routes. The proposed system is based on a multiagent architecture, all agents will work together to monitoring the traffic road congestion and help drivers to achieves their destination by following the best routes with less congestion.

Keywords: Ant colony system · Traffic monitoring system ·
Multi-agent system · Vehicle routing problem ·
Distributed swarm intelligence

1 Introduction

The traffic congestion is a serious problem [1] which is not given any indications of improvement. Recently, various solutions have been offered to reduce the traffic congestion problem. Those strategies based on several technologies [8], some of them are based on GPS coordinates and satellites or using cellular data and

© Springer Nature Switzerland AG 2019
F. Khoukhi et al. (Eds.): AIT2S 2018, LNNS 66, pp. 230–237, 2019.
https://doi.org/10.1007/978-3-030-11914-0_25

GSM, other one use image processing and CCTV cameras, some solution are based on RFID (radio frequency identification). Recently, some works have been done based on vehicular network infrastructure which aims to create a network of communication between the different entities that collaborate in the traffic network such as vehicles and light signs and intersections, etc. This new concept is based on the vehicular ad-hoc network (VANET) [5,6] which is one of the famous technologies implemented in the intelligent transportation system (ITS) concepts.

In this work, we introduce a new distributed vehicle traffic routing system (DVTRS) [3,7,9] based on a multi-agent architecture. The system is based on a collective group of agents, they cooperate together for collecting traffic data and proposing the best routes with less traffic jam to the driver of the vehicle. They work in parallel to solve the complex problem and find optimal solutions. The system is based on two main components: (i) the VANET technology for preparing the infrastructure of the road network which uses several entities that guarantee the best exchange of information between vehicles and road controller agents. (ii) The Ant colony optimization (ACO) [1,11] which is used to compute the shortest path based on the traffic data collected from the road network. The traffic information is collected from several sensors implemented at each intersection that are able to define the traffic flow on those segment of routes.

The remainder of this work is classified as follows. The next section reviews related work. In Sect. 2 a literature and a background review are described. Section 3 describe the concept of the VANET architecture. An overview of the ant colony optimization techniques (ACO) is presented in Sect. 4. The next section outlines the concept of the combining the technique of the ACO with the technology of VANET. In Sect. 5 the proposed distributed strategy is described in detail. Finally, the conclusion is given in Sect. 6.

2   Related Work

In this work [5], authors propose a novel routing protocol named as Intersection-based Delay sensitive Routing using Ant colony optimization (IDRA). In this strategy, they determine a mathematical delay pattern for a road section. Then using the intersection concept, the proposed system will use the ACO technique to find the optimal route with the minimum travel time. Each deployed ant will try to explore the road network by visiting a set of intersections, each route will be selected according to a local delay or a global delay from the current location toward the final intersection.

[10] present a bio-inspired method based on the Ant Colony Optimization techniques to find the optimal path in a given road network. This method is based on two constraints such as the delay time and availability of the route to find the best routes. The proposed solution is applied to a dynamic environment and all the new changes are taken into consideration during the process of the optimization done by the ACO.

A new clustering method has been introduced in [12] that combines ant colony optimization (ACO) with the genetic algorithm (GA) for the VANET road network to increase the flexibility and the lifetime of the road network. The ACO is implemented to search for the optimal solution that aims to reduce the total number of clusters in the road network then the GA is employed to search for the best cluster centers.

3   Vehicular Ad-hoc Network (VANET) Infrastructure

With the revolution of technologies, modern cars are equipped with various intelligence components such as OBD-II system, GPS device, emergency services and self-driven system. However, we need to find an optimal approach that help us to assign some routes for specific vehicles or give to drivers information about the congested routes in the road network. The object of this study is to construct an intelligent traffic monitoring system using VANET combined with the ACO [2,7] to detect traffic jam in real-time and guiding the vehicle to follow specific routes with less congestion.

Vehicular ad-hoc network (VANET) [4,6] is a particular form of Mobile ad-hoc network (MANET) which allow us to build a communication network made from a vehicle to vehicle or a roadside units to vehicles. VANET is characterized by his self-organizing and autonomous, where entities in VANET can work as clients and (or) servers for exchanging and sharing traffic road information, Fig. 1 shows in detail the architecture of the VANET. The technology of VANET has taken large attention from academic research and government and industry. VANET has been implemented in various strategies such as Security distance warning, Vehicle collision warning, Driver assistance, Internet access, Cooperative driving, Automatic parking and Dissemination of road information.

The VANET architecture is based on three main entities: A Monitoring center (MC) installed at each section which contains a local database used for storing and analyzing data received from both RSU (Road Side Unit) and On-Board vehicle unit (OBU). The MC can interact directly with each RSU to fetch information about the state of the road and get the number of vehicles that are on this route at real-time. On the other hand, the MC will help the OBU to achieve his destination by giving him the optimal path according to the information collected from the set of the RSU's in this area. The RSU is defined to collect information from routes such as the number of the vehicle and the traffic flow, those data will be sent periodically to the MC to help them in the process of finding the shortest path for vehicles. The last entity is named as OBU, this entity will control the state of the vehicle and will communicate with the MC and the RSU using the internet to send a request and handle responses.

4   Combine the ACO Technique with the VANET Architecture

The road network will be viewed as a graph made from a set of intersections (nodes) and routes (edges). This graph will be represented as follow: $G = (I, R)$

Fig. 1. Overall VANET infrastructure

Where I=$(I_1, I_2, ..., I_n)$ is the set of intersections and R is the set of routes. The main objective of this architecture is guiding vehicles to reach their destination in the fastest time while avoiding the traffic jam on the road network. Avoiding traffic jam is done by skipping routes with a higher number of non-moving vehicles, traffic flow delays in this instance are based only on the congestion. Each R_{ij} is identified by the distance d_{ij} and traffic flow T_{ij}. The time consumed to traverse R_{ij} is independent of the time consumed to traverse other routes because each route has its number of non-moving vehicles controlled by the RSU. Therefore the total traveling time for crossing a route will be the sum of all time taken to visit each intersection.

The final goal is to find the optimal solution that minimizes the two constraints such as the distance and the traveling time. So in this problem, we will try to use two parameters (distance, traffic flow) to find the best solution to this problem. The vehicle will apply a probabilistic transition to move from an intersection I_i toward I_j according to the rule below:

$$p_{ij}^m(it) = \frac{T_{ij}(it)^\alpha * d_{ij}(it)^\beta}{\sum_{l \in \omega} T_{il}(it)^\alpha * d_{il}(it)^\beta} \tag{1}$$

where the parameter (α, β) control the influence of the traffic flow (pheromone) and the distance on the prepared solution, the parameter ω is the set of the non-visited intersections. The level of the traffic flow (pheromone) will evaporate (ρ rate of evaporation) depending on the information received from the RSU (the rate of the non-moving vehicles on this route). The traffic flow update is done according to this equation:

$$T_{ij} = (1 - \rho).T_{ij} + \Delta_{ij} \tag{2}$$

The argument Δ_{ij} is related to the driver to see if the vehicle has already visited this route R_{ij} or not. The new amount of traffic added on this route can be computed as follows:

$$\Delta_{ij} = \sum_{n=1}^{N} \frac{R_{ij}^k}{t_m(k)}$$

where $tm_{(k)}$ represents the time needed to cross the distance d_{ij} of the route R_{ij} by the vehicle k based on the speed (ω_k) allowed on this route and the number of the available non-moving cars on this route.

$$tm_{(k)} = \frac{d_{ij}}{\omega(k)}$$

So at each intersection, the vehicle will try to select the next intersection to visit based on the (Eq. 1) and select this route based on two parameters the distance and the traffic flow on this route. The system will repeat the same behaviour until the destination position is reached or the stop condition is met. At the end, the system will show the best routes that exist between these two locations, for each route they will give us the total travel time consumed to cross this route with the total length (km) of this trip. The (Eq. 1) can be transformed into:

$$p_{ij}^m(it) = \frac{T_{ij}(it)^\alpha * d_{ij}(it)^{1-\alpha}}{\sum_{l \in \omega} T_{il}(it)^\alpha * d_{il}(it)^{1-\alpha}} \tag{3}$$

5   A Distributed Strategy for Solving the Routing System Problem

In this section, we will introduce the distributed strategy followed by this solution to find the optimal path that avoids the traffic jam. Our architecture will be based on three main agents (Fig. 2): MA (Monitoring Agent), VA (Vehicle Agent), (RA) Routing agent. The system will work in a decentralized environment by running these agents in parallel and trying to find useful solutions in the minimum time.

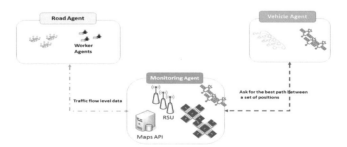

Fig. 2. Decentralized architecture for solving traffic congestion based on multi-agent architecture

5.1 Monitoring Agent (MA)

The MA will control the workflow of our architecture by handling the request from the vehicle (VA) and give him the best path that will be followed to avoid traffic jam, on the other hand, it will work with the set of the existing RSU's in the area to collect all the data about the state of the route at each intersection. When a driver needs to go toward a specific destination, the MA will create a graph made from a set of intersections by calling the Google Maps API services, this open source API will help as to draw this map and generate a set of intersections between the two given locations (start, end). For the cost of each route, the MA will take the number of vehicles presented by the traffic flow on this route as the value of the link between two adjacent intersections and the linear distance between those positions. This value will be fetched from the list of the RSU that control these routes. Based on that information, the MA will be able to define the road network that will be sent to the RA. In the next step, the RA will apply the AS algorithm [3,7,9] to find the best path that exists between the start and the destination, those paths will not be the shortest in terms of distance but also in terms of traffic flow. The RA will search for the best routes by creating a set of artificial ants named as worker agents (WA), each one will take a route and will try to visit a set of intersections based on the probabilistic transition defined in the AS algorithm. To select the next transition we will use the traffic flow on this route to see the best routes that have less traffic jam (less number of non-moving vehicles). (Figure 3) describe in detail the main behavior of the MA in this architecture.

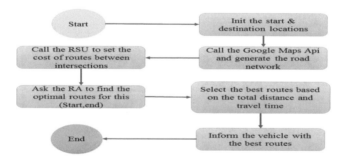

Fig. 3. Simulation flow diagram for monitoring traffic congestion using VANET strategy

5.2 Vehicle Agent (VA)

In the proposed solution we are based on a road graph made from a set of locations defined by two attributes (latitude, longitude) and to control our vehicle we need to know the start and the destination of this vehicle based on here GPS location. The VA is defined to control the state of the vehicle by sharing a set of information about this vehicle with the MA, such as his current speed, actual GPS position, etc. When the optimal path is produced by the MC, this agent will help the driver to complete his journey by following this proposed path planning.

5.3 Routing Agent

To compute the best route for the driver we are based on the RA (Fig. 4), this agent uses the techniques of the Ant System (AS) Algorithm to find the optimal trip that exists on the given map. In the beginning, it will handle the start node and the destination node of the driver and try to find the best trip that can exist by visiting a set of intersections. To make this job, our architecture will use a set of artificial ants named as worker agent (WA), each WA will try to visit all the available intersection by applying the probabilistic transition rule defined by the AS algorithm and select the next intersection to visit based on the number of the non-moving vehicles and the length of the route. So here we are based on the traffic flow to prepare our trip and not only finding the shortest route in terms of the distance. Secondly, when the WA finished his tour by arriving at the destination node, it will be able to apply a local update of information using the local update of pheromone rule on the list of the routes already visited. The created route will be sent to the RA to compare them with the list of the solutions proposed by the other WA. To select the best route, the RA will take the route with less traffic flow that means routes with the minimum number of non-moving vehicles. In the end, the optimal route will be sent to the MC and the driver can start his journey by following this proposed path.

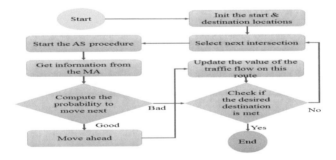

Fig. 4. Simulation flow diagram for vehicle routing system

6 Conclusion

This work presents a new intelligent traffic routing controlling system that can effectively reduce and avoid the traffic congestion on the routes of drivers. The proposed strategy is based on the VANET architecture combined with a distributed ant system algorithm (DAS). The DAS is an instance of swarm intelligence methods that show better results in the domain of finding shortest and optimal routes. The whole system is based on other technologies to make the process easy to work and to propose good result that will be used by the driver to avoid or to skip routes with higher traffic congestion. The simulation feedback shows that the proposed method not only decrease traffic congestion by re-routing vehicles toward routes with less traffic but also collaborate in reducing the total waiting time when the traffic congestion occurs on specific routes.

References

1. Deshmukh, A.R., Dorle, S.S.: Bio-inspired optimization algorithms for improvement of vehicle routing problems. In: 7th International Conference on Emerging Trends in Engineering & Technology (ICETET) (2015)
2. Neto, A.F., Cardoso, P.A.: Dynamic vehicle programming and routing system applied to wheelchair transportation. IEEE Lat. Am. Trans. **15**(2) (2017)
3. Kaplar, A., Vidakovic, M., Luburi, N., Ivanovic, M.: Improving a distributed agent-based ant colony optimization for solving traveling salesman problem. In: The 40th International Convention on Information and Communication Technology, Electronics and Microelectronics (MIPRO), Opatija, pp. 1144–1148 (2017)
4. Raut, C.M., Devane, S.R.: Intelligent transportation system for smartcity using VANET. In: International Conference on Communication and Signal Processing (ICCSP) (2017)
5. Li, G., Boukhatem, L.: An intersection-based delay sensitive routing for VANETs using ACO algorithm. In: 23rd International Conference on Computer Communication and Networks (ICCCN) (2014)
6. Al Najada, H., Mahgoub. I.: Anticipation and alert system of congestion and accidents in VANET using big data analysis for intelligent transportation systems. In: IEEE Symposium Series on Computational Intelligence (SSCI) (2016)
7. Elgarej, M., Mansouri, K., Youssfi, M., Benmoussa, N., elfazazi, H.: Distributed swarm optimization modeling for waste collection vehicle routing problem. Int. J. Adv. Comput. Sci. Appl. (IJACSA)
8. Kimura, M., et al.: A novel method based on VANET for alleviating traffic congestion in urban transportations. In: IEEE Eleventh International Symposium on Autonomous Decentralized Systems (ISADS) (2013)
9. Kromer, P., Gajdo, P., Zelinka, I.: Towards a network interpretation of agent interaction in ant colony optimization. The IEEE Symposium Series on Computational Intelligence, Cape Town, pp. 1126–1132 (2015)
10. Majumdar, S., et al.: An efficient routing algorithm based on ant colony optimisation for VANETs. In: IEEE International Conference on Recent Trends in Electronics, Information & Communication Technology (RTEICT) (2016)
11. Mahalingam, V., Agrawal, A.: Learning agents based intelligent transport and routing systems for autonomous vehicles and their respective vehicle control systems based on model predictive control (MPC). In: IEEE International Conference on Recent Trends in Electronics, Information & Communication Technology (RTE-ICT) (2016)
12. Goswami, V., Verma, S.K., Singh, V.: A novel hybrid GA-ACO based clustering algorithm for VANET. In: 3rd International Conference on Advances in Computing, Communication & Automation (ICACCA) (Fall) (2017)

Intelligent Information Systems
and Modeling

New Approach for Solving Infinite Cycles Problem During Modeling

Abdessamad Jarrar[(✉)], Taoufiq Gadi, and Youssef Balouki

Faculty of Sciences and Technologies SETTAT, Computing,
Imaging and Modeling of Complex Systems Laboratory, Settat, Morocco
abdessamad.jarrar@gmail.com

Abstract. When modeling a system using the formal method Event-B, One of the most common problems that may occur during a system processing is infinite cycles. Infinite cycles mean that only certain events are allowed to be executed. In this paper, we present a method to verify the absence of unexecuted events in future system basing on graph theory. This type of verification is not treated by Event-B this mean that even if your system is verified by proof obligations your system is not ensured to occur correctly. In a simple system, it may be very easy to verify that all the events will be executed but in complex system with so many events, it will be impossible. Our method is a verification method based on strongly connected digraph (directed graph), this method verifies the execution of all events in a system and this is why we will call a system verified by our method a strongly connected system.

Keywords: Event-B · Infinite cycle · Graph theory ·
Strongly connected digraph

1 Introduction

Event-B is a formal method used to model systems especially control systems and distributed systems. Event-B is based on a mathematical language; this allows us to prove using a mathematical logic that our system is correct; these proofs are called proof obligations. Proof obligations in event-B verify the correctness of system logic. In other term, the consistency of the system, it also verifies the absence of deadlocks. However, these proofs do not treat the case of infinite cycles and unexecuted events. In this paper, we present a method to treat this case basing on graph theory.

In the beginning, we start by presenting mathematically our method and how we will use graphs to represent a system modeled in event-B. Then, we will propose an algorithm to verify that a certain system is strongly connected. We will also present an example of simple system that is proved by event-B standard proof obligations; despite, it has unexecuted events within, and then we will apply our method to prove that the system is not strongly connected.

© Springer Nature Switzerland AG 2019
F. Khoukhi et al. (Eds.): AIT2S 2018, LNNS 66, pp. 241–248, 2019.
https://doi.org/10.1007/978-3-030-11914-0_26

2 Modeling in Event-B

Due to the high risk in safety-critical systems, it is highly recommended to base their engineering on a certain theory. For example, electrical engineering is based on Maxwell's equations and Kirchhoff's laws; civil engineering is based on geometry and theory of material's strength. Similarly, software engineering has formal methods which are less considered. Therefore, there are engineers who do not know any theory building software, and the software often has bugs that may cost millions to fix. Now, formal methods can be used for verifying and specifying software in order to highly guarantee bugs' absence.

Event-B is a formal method that provides the correct-by-construct approach and formal verification by theorem proving [2, 14]. Models in Event-B are presented based on abstract state machine notion, which presents the model states in term of a set of variables; these states are constrained by invariants. Invariants are the necessary properties that must be preserved during system function. Statuses transitions are described by events, which are a set of actions. Each action changes the value of certain variable. Events may have some necessary conditions to be triggered; these conditions are called guards. Models in Event-B include sets, constants and axioms representing the static part of the model.

One of the main features of Event-B is refinement, which means starting modeling with an abstract model and then enriches it in successive steps by adding more details. This techniques makes modeling easier than trying to model the whole system at once, we focus on a limited number of requirement in each step under the condition of cleverly choose the refinement strategy. The figure below presents an outline of the process of modeling in Event-B (Fig. 1):

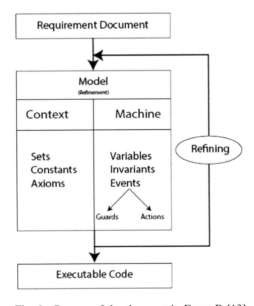

Fig. 1. Process of development in Event-B [13]

Event-B models consistency, invariant preservation and the correctness refinement–refinements should not contradict- are ensured by discharging a number of verification conditions called proof obligations. For example, to prove that an invariant is preserved by an event, we prove that if an invariant is preserved before the event it will remain preserved after it. Mathematically speaking, let I be the model invariant, A are axioms, c are constants, s are set, v are variables before the event occurrence and v′ are variables after the event. The following logical formula should be proved in order to prove invariant preservation:

$$A(c,s) \wedge I(v,c,s) \vdash I'(v',c,s) \tag{1}$$

Most of proof obligations are discharged automatically by means of a platform called Rodin. The remaining proofs may be dealt using an interactive prover included in Rodin.

3   Mathematical Specification of the Proposed Method

Event-B is an established formalism that has been applied to a range of systems especially control systems and distributed systems [1]. A development in event-B is a set of machines and contexts, a context is the static part of the developed model while the machine is the dynamic one. This development is based on refinement; refinement is creating an abstract model and then enriching it in successive steps [2]. This method of refinement makes modeling easier than trying to model the whole system at once.

In this chapter, we present how to transform a system modeled in event-B into a graph. Then, we present our verification method that will be applied on the resulting digraph. As we are interested in the dynamic part of the system we will be interested in only the *machine*. Also, we apply our method on every refinement after doing proofs obligation, whereby the resulting system is correct by construction [3].

A graph is a pair G = *(V, E)*, where V is the vertex set and *E* is a directed edges set [4]. Usually a graph is presented as follow (Fig. 2):

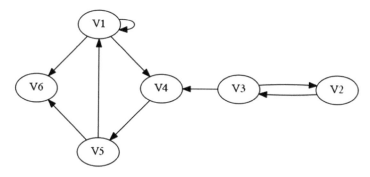

Fig. 2. Example of a digraph

To transform an event-B model into a graph, we define the set of vertex and the set of directed edges. The set of vertex represent the events except the initialization event, this method is similar to the method used in Petri net [5, 6]. Let $e_1, e_2 \ldots e_n$ the events of a system and *Init* is the initialization event and *Events* is the set of all the events, then the set of vertex is the following:

$$V = \{e_1, e_2, ..e_n\} = Events \backslash \{Init\}$$

An edge is a finite set $E(G)$ of distinct ordered pairs of distinct elements of $V(G)$ [7]. In our case, a vertex e_i is related to another vertex e_j if the after-predicate of the event e_i may verify the guards of e_j. We present this relation as follow:

$$e_i R e_j \Leftrightarrow the\ after\ predicate\ of\ e_i\ is\ verifying\ the\ guards\ of\ e_j$$

This means that the set of edges can be presented as follow:

$$E = \{(x, y) | x R y\}$$

To verify that a system is strongly connected, we have to prove that the representation digraph is strongly connected, this mean that we can reach any vertex starting from any other vertex by traversing a certain path (a set of successive directed edges). While the vertex are the events of the system, Verifying the connection of digraph ensure that all the events may occur no matter what event is executed, this ensure the absence of infinite cycle and unexecuted events. Mathematically this property is the following:

$$(\forall x, y \in V)\ there\ is\ a\ path\ from\ x\ to\ y \Leftrightarrow (\forall x, y \in V)(\exists x_1, x_2, ..x_n \in V)$$
$$: xRx_1, x_1Rx_2 ..\ and\ x_n\ Ry$$

In the case of a direct edge from x to y the set of x_i is $\{x\}$ or $\{y\}$ themselves.

4   An Overview of the Problematic of Infinite Cycles in Event-B Digraph

In this chapter, we give a brief overview about the problematic with an example. As mentioned before, an infinite cycle is a sequence of events that execute endlessly; this means that we will have some events that are not in the cycle, these events will never be executed when entering the cycle [8]. Here is an example of an infinite cycle (Fig. 3):

In the graph above, if we start with executing the event V_1, the event V_4 will never be executed; this mean that there is no reason for its existence, this is why need to verify the absence of infinite cycles and unexecuted events in the resulting digraph of an event-B model, This means that this verification is highly recommended [9]. Event if we start with executing V_4, we still have a problem, when V_1 or V_3 occur, V_4 will never occur again. In some cases, we may have a system in which an event must be

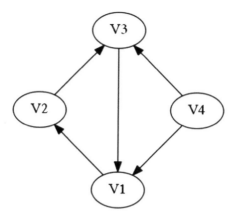

Fig. 3. Example of an infinite cycle

executed only once, and in this case we remove this event from the list of graph events, then, we verify that the resulting digraph is strongly connected.

5 Verification Algorithm

Now that we presented how to transform an event-B model into a digraph, we apply an algorithm to verify that our graph is strongly connected. To do so, we will use an algorithm based on Kosaraju's algorithm [10–12]:

1. Let $G = (V, E)$ where V is the set of vertices and E is the set of edges.
2. Let the variable *isConnected* := *True*.
3. Remove all vertex $V1$ representing events that have to be executed only once.
4. Initialize an ordered list of vertices L as an empty list. This list will define the order in which we will proceed our algorithm
5. For every vertex v in our graph, if v is not visited mark as visited and for each out-neighbor of v do 2, then add v to L.
6. Initialize a list of vertices $SCC(l)$ (Strongly Connected Components of V) as the Singleton $\{l\}$, where l is the last vertex in L.
7. For each in-neighbor vertex W of l, if W is not assigned to a root then assign W to the root l and add W to $SCC(l)$, then repeat 5 for each in-neighbor of W.
8. If $SCC(l) \neq V\backslash V1$ then *isConneted* := *false*.

The variable *isConnected* determine the system is strongly connected or not.

6 Example of Infinite Cycle Problem

Here, we will propose an abstract example of a system digraph representation. In this example, we have unexecuted events and an infinite cycle; we will apply our method to prove that this system is not strongly connected.

Let the digraph below the representation of a certain system (Fig. 4):

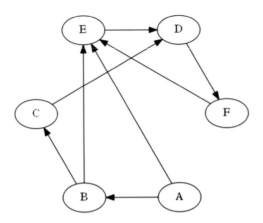

Fig. 4. Digraph of a certain system

Now we can apply our algorithm as follow:

1. The graph above is a graph G = (V, E) where:

$$V = \{A, B, C, D, E, F\}$$

$$E = \{(A, B), (A, E), (B, E), (B, C), (C, D), (D, F), (F, E), (F, E), (E, D)\}$$

2. Let isConnected := True
3. For this step we suppose that we want the event represented by the vertex A to be executed only once, due to this, we must remove A from the studied graph:

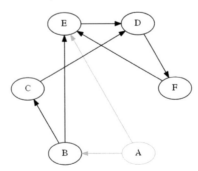

4. Initializing the ordered list L as an empty list:

$$L = \phi$$

5. In the beginning we chose a random vertex to start with, in this example we start with E. In the end of this step L will be:

$$L = \{F, D, E, C, B\}$$

6. Since A is the last vertex of L, SCC(B) will be initialized as follow:

$$SCC(B) = \{B\}$$

7. Since there is no in-neighbor of B the set SCC(B) will not change during step 7.
8. While SCC(B) = {B} ≠ E\{A} then the system is not strongly connected and isConnected := false.

7   Conclusion

We have presented an algorithm verify the existence of unexecuted events and infinite cycles. To avoid this, we propose an algorithm to verify that the digraph representing a system is strongly connected. To transform our system modeled in event-B into a digraph, we represent events by vertices and the edges linking two events if the after-predicates of an event may verify the guards of the other. After that, we apply our algorithm that is based on kosaraju's algorithm to verify that our digraph is strongly connected. For future works, we are working on developing other methods for system correctness verification beside event-B proof obligations; also we're developing a complete modelization of the aeronautic system.

References

1. Butler, M., Abrial, J.R, Banach, R.: Modelling and refining hybrid systems in Event-B and Rodin, chap. 3, pp. 29–42 (2016)
2. Abrial, J.-R.: Modeling in Event-B: System and Software Design. Cambridge University Press, Cambridge (2008). 586 p.
3. Hoang, T.S., Kuruma, H., Basin, D., Abrial, J.-R.: Developing topology discovery in Event-B (2009). 19 p.
4. Ruohonen, K.: Graph Theory. Translation by Janne Tamminen, Kung-Chung Lee and Robert Piché (2013). 110 p.
5. Peterson, J.L.: Petri Nets. Department of Computer Sciences, The University of Texas, Austin, Texas 78712 (1977)
6. Geeraerts, G.: An Introduction to Petri nets and how to analyse them…, Groupe de Vérification Département d'Informatique Université Libre de Bruxelles, 165 p.
7. Wilson, R.J.: Introduction to Graph Theory, 4th edn. Longman Group Ltd. (1998), 171 p. ISBN 0-582-24993-7
8. Diestel, R., Kuhn, D.: On infinite cycles II. Combinatorica **24**(1), 91–116 (2004)
9. Bruhn, H., Stein, M.: On end degrees and infinite cycles in locally finite graphs. Combinatorica **27**(3), 269–291 (2007)

10. Sharir, M.: A strong connectivity algorithm and its applications to data flow analysis. Comput. Math Appl. **7**, 67–72 (1981)
11. Cormen, T.H., Leiserson, C.E., Rivest, R.L., Stein, C.: Introduction to Algorithms, 3rd edn. The MIT Press, Cambridge (2009). ISBN 0-262-03384-4
12. Aho, A.V., Hopcroft, J.E., Ullman, J.D.: Data Structures and Algorithms. Addison-Wesley, Boston (1983)
13. Jarrar, A., Balouki, Y.: Formal modeling of a complex adaptive air traffic control system. Complex Adapt. Syst. Model. **6**(1), 6 (2018)
14. Vistbakka, I., Troubitsyna, E.: Towards integrated modelling of dynamic access control with UML and Event-B. arXiv preprint arXiv:1805.05521 (2018)

P_a-QUACQ: Algorithm for Constraint Acquisition System

Hajar Ait Addi$^{(\boxtimes)}$ and Redouane Ezzahir

LISTI/ENSA, University of Ibn Zohr, Agadir, Morocco
{hajar.aitaddi.edu,r.ezzahir}@uiz.ac.ma

Abstract. Constraint acquisition system assists a non-expert user in modeling her problem as a constraint network. T-QUACQ is a recent learner system that acquires constraint network by generating queries in an amount of time not exceeding a waiting time upper bound. The hindrance is the risk of reaching a premature convergence state. In this paper we present P_a-QUACQ, a new algorithm based on T-QUACQ. P_a-QUACQ is able to made T-QUACQ more efficient in terms of time and convergence. Finally, we give an experimental evaluation of our algorithm on some benchmarks.

Keywords: Constraint programming · Constraint acquisition · QUACQ

1 Introduction

Constraint programming (CP) is a powerful paradigm to model and solve complex combinatorial problems. Nevertheless, the modelling task, that aims to encode a problem as a constraints network, requires some expertise in constraint programming. Several learning algorithms have been developed to facilitate access to the constraint technology by novices. The matchmaker agent [11] proposed by Freuder and Wallance. When the system propose an incorrect solution, the agent asks the user to communicate a new constraint that explains why she considers a proposed solution as a wrong one. Lallouet et al. introduced a system based on inductive logic programming [8]. Beldiceanu and Simonis have proposed MODELSEEKER, a system devoted to problems with regular structures and based on the global constraint catalog [7]. Bessiere et al. proposed CONACQ, which interactively proposes to the user membership queries (i.e., complete examples) to be classified by the user [2,10].

Bessiere et al. proposed QUACQ (for Quick Acquisition), an active learning system that is able to ask the user to classify partial queries [3,5]. If the user says *yes*, QUACQ removes from the search space all constraints violated by the positive example. If the user says *no*, QUACQ finds the scope of one of the violated constraints. Arcangioli et al. proposed the MULTIACQ system [4]. Instead of finding the scope of one constraint, MULTIACQ reports all the scopes

© Springer Nature Switzerland AG 2019
F. Khoukhi et al. (Eds.): AIT2S 2018, LNNS 66, pp. 249–256, 2019.
https://doi.org/10.1007/978-3-030-11914-0_27

of constraints of the target network violated by the negative example. Hajar et al. proposed T-QUACQ [1], as an attempt to make QUACQ more efficient in terms of time-consuming between queries. T-QUACQ generates queries in an amount of time not exceeding a waiting time upper bound. A major bottleneck of T-QUACQ is the risk of reaching a premature convergence state.

In this paper, we introduce a new approach P_a-QUACQ to allow T-QUACQ to converge. Instead of returning premature convergence, we force T-QUACQ to perform as QUACQ. In other words, when the time bound is reached, our new algorithm P_a-QUACQ tries to find a query without taking into consideration the time cutoff. We experimentally evaluated our approach on several benchmark problems. The results show that P_a-QUACQ improves the basic T-QUACQ in terms of convergence state without affecting negatively the time of generating queries.

The rest of this paper is organized as follows. Section 2 presents the necessary background on constraint acquisition. Section 3 describes P_a-QUACQ algorithm. Section 4 reports the experimental results we obtained when comparing P_a-QUACQ to T-QUACQ. Section 5 concludes the paper.

2   Background

The constraint acquisition process can be seen as an interplay between the user and the learner. User and learner need to share a *vocabulary* to communicate. A vocabulary is a set of n variables $X = \{x_1, \ldots, x_n\}$ and a domain $D = \{D(x_1), \ldots, D(x_n)\}$, where $D(x_i) \subset \mathbb{Z}$ is the finite set of values for x_i. A constraint c_Y is defined by a sequence Y of variables of X, called *the constraint scope*, and the relation c over D of *arity* $|Y|$. An assignment e_Y on a set of variables $Y \subseteq X$ *violates* a constraint c_Z (or e_Y is *rejected* by c_Z) if $Z \subseteq Y$ and the projection e_Z of e_Y on the variables in Z is not in c. A *constraint network* is a set C of constraints on the vocabulary (X, D). An assignment on X is a *solution* of C if and only if it does not violate any constraint in C. $sol(C)$ represents the set of solutions of C.

In addition to the vocabulary, the learner owns a *language* Γ of relations, from which it can build constraints on specified sets of variables. from the constraint language Γ on the vocabulary (X, D), from which the learner builds the constraint network. We denote by $B[Y]$ the set of all constraints c_Z in B, where $Z \subseteq Y$. The *target network* is a network C_T such that for any example $e \in D^X = \Pi_{x_i \in X} D(x_i)$, e is a solution of C_T if and only if e is a solution of the problem that the user has in mind.

Given an example e (resp e_Y) and a partial network learned so far. We denote e by $e \models (C_L \wedge \neg B)$ (resp $e_Y \models (C_L[Y] \wedge \neg B[Y])$) if e (e_Y) does not violate any constraint in the current learned network C_L (resp $C_L[Y]$), and it violates at least one constraint of the current bias B (resp $B[Y]$).

A *membership query* ASK(e) is a classification question asked to the user, where e is a *complete* assignment in D^X. The answer to ASK(e) is *yes* if and only if $e \in sol(C_T)$. A *partial query* ASK(e_Y), with $Y \subseteq X$, is a classification question asked to the user, where e_Y is a *partial* assignment in $D^Y = \Pi_{x_i \in Y} D(x_i)$. The answer to $\text{ASK}(e_Y)$ is *yes* if and only if e_Y does not violate any constraint in C_T. A classified assignment e_Y is called a positive or negative *example* depending on whether $\text{ASK}(e_Y)$ is *yes* or *no*. For any assignment e_Y on Y, $\kappa_B(e_Y)$ denotes the set of all constraints in B rejecting e_Y.

We now define *convergence*. Given a set E of (complete/partial) examples labeled by the user as positive or negative. We say that a constraint network C *agrees with* E if positive examples in E do not violate any constraint in C, and negatif examples violate at least one constraint in C. The learning process has *converged* on the learned network $C_L \subseteq B$ if:

1. C_L agrees with E,
2. For any other network $C' \subseteq B$ agreeing with E, we have $sol(C') = sol(C_L)$.

The learning process reached a *premature convergence*, if only (1) is approved. If there is no existing C_L, we have collapsed. This happens when C_L can't be expressed by B.

3 P_a-QUACQ

3.1   Description

In this section, we present a description of P_a-QUACQ algorithm (for partial QUACQ). P_a-QUACQ takes as input a bias B and returns a constraint network C_L. The main difference between P_a-QUACQ and T-QUACQ, is the fact that T-QUACQ stop learning when it couldn't find a query on subset $Y \in X$ in a bounded time time_bound (case of *premature convergence*), whereas P_a-QUACQ tries first to generate a query on subset $Y \in X$ in an amount of time not exceeding a waiting time upper bound, and when it fails, he performs like QUACQ on terms of generating queries: P_a-QUACQ attempts to generate a query on the whole set X. We could say that P_a-QUACQ is an intermediate algorithm between QUACQ and T-QUACQ.

The algorithm P_a-QUACQ (see Algorithm 1) takes as arguments the bias B, the set of constraints C_L learned so far, a time threshold τ, a time bound time_bound, a reduction factor $\beta \in]0, 1[$, and a query size ℓ.

When we enter the main loop, we set the boolean *noSolution* to *false* (line 4). This boolean indicates whatever our algorithm should continue to search for a query or not. In line 5, we initialize the counter *time* to zero. *time* counts the consumed time during the execution of the second loop (block: 7–20). In line 8, we ensure that the query size ℓ will be always greater or equal to the smallest arty in B. In 10 line, we select a subset of variables Y of size ℓ with respect that $B[Y]$ is not empty. Next, we enter a loop to find a query.

If *time* does not exceed time_bound or *noSolution* doesn't equals to *true* or the bias B isn't empty, then we attempt to generate a query e on Y of size ℓ in a time t less than τ (line 11). If such query exists, we set *noSolution* to *true* to exit the loop. If such example doesn't exist and the time cutoff τ wasn't

Algorithm 1. P_a-QUACQ

```
 1  C_L ← ∅ ;
 2  initialize(β, τ, ℓ, time_bound) ;
 3  while true do
 4  │   noSolution ← false ;
 5  │   time ← 0 ;
 6  │   e ← nil ;
 7  │   while time < time_bound ∨ noSolution ∨ |B| ≠ ∅  do
 8  │   │   ℓ ← max(ℓ, minArity(B)) ;
 9  │   │   t ← min(τ, time_bound − time) ;
10  │   │   choose Y ⊆ X s.t. |Y| = ℓ ∧ B[Y] ≠ ∅ ;
11  │   │   e ← solve(C_L[Y] ∧ ¬B[Y]) in t < τ ;
12  │   │   if e ≠ nil then
13  │   │   │   noSultion = true;
14  │   │   else
15  │   │   │   if t < τ then
16  │   │   │   │   C_L ← C_L[Y] ∪ B[Y] ;
17  │   │   │   │   B ← B \ B[Y] ;
18  │   │   │   else
19  │   │   │   │   time ← t + time ;
20  │   │   │   │   ℓ ← ⌊β · ℓ⌋ ;
21  │   if !noSolution then e ← solve(C_L ∧ ¬B);
22  │   if e = nil then return"Convergence";
23  │   else
24  │   │   if Ask(e) = yes then
25  │   │   │   B ← B \ κ_B(e) ;
26  │   │   │   ℓ ← min(⌈ℓ/β⌉, |X|) ;
27  │   │   else
28  │   │   │   c ← FindC(e, FindScope(e, ∅, X, false)) ;
29  │   │   │   if c ≠ nil then C_L ← C_L ∪ {c} ;
30  │   │   │   else return"collapse" ;
31  │   │   │   ℓ ← ⌊β · ℓ⌋;
```

reached, this means that $C_L[Y] \wedge \neg B[Y]$ is unsatisfiable. So, the constraints in $B[Y]$ are redundant to C_L, and it will be removed from B and added to C_L (line 16). If such query doesn't exist and the time τ was reached, so the problem $C_L[Y] \wedge \neg B[Y]$ is too hard to be solved in τ. Then, we reduce the query size ℓ by the factor β and we increment the counter $time$ by t (lines 19, 20).

If we have exceed the time_bound and $noSolution$ still equals to $false$, this means that we have failed to generate a query in the time limit time_bound. Then our algorithm performs as QUACQ (line 21). We try to generate a query e on X.

After generating a query, we have converged if the returned query doesn't exist. Otherwise, we propose the example e to the user to classify it. If the user says "yes", we remove all constraint that reject e from B. We also increase ℓ by the factor β (line 26). If the user says"no", we are sure that e violates at least one constraint of the target network C_T. Then, we call the functions `FindScope` and `FindC` (line 28) to find a voilated constraint. If a constraint c is returned, we add it to the learned network C_L and we decrease the query size ℓ by the factor β (lines 29, 31). Otherwise, we could not find in B a constraint that reject the negative example. So, we return "collapse" in line 30.

3.2   Analysis

Proposition 1 (Termination). *The second loop from line 7 to line 20 is guaranteed to terminate if* `time_bound` $< \infty$.

Proof. This loop terminates when $time >$ `time_bound` or *noSolution* is *true*. If $e \neq nil$, it is trivial. Suppose now that e doesn't exist ($e = nil$), `time_bound` $< \infty$. At each iteration, we will increase $time$. After a finite number of iteration, the value of $time$ will be greater than the `time_bound`. As a result, the loop will terminate.   □

Note 1. *Throughout this article, the* `time_bound` *is a finite number in order to ensure the termination of our algorithm.*

Proposition 2 (Convergence). *if* $C_T \subseteq B$ P_a-QUACQ *is guaranteed to converge.*

Proof. The proof is by contradiction. Let suppose that $C_T \subseteq B$ and P_a-QUACQ collapses. This happens when P_a-QUACQ could not find a constraint c in B to explain a negative example e (line 29). This means that c does not exist in B, or c was removed during the learning process. However, $C_T \subseteq B$; so c is in the B. Let prove that the removal of c is contradictory. Let assume that c was removed during the learning process (line 25). Thus, c voilates a positive example. Contradiction!   □

4   Experimental Evaluation

This section presents the experiments that we made to evaluate the performance of our approach. All tests were performed using the Choco solver[1] version 4.0.4 on an Intel Core i7-7500U CPU @ 3.50GHz with 8 GB of RAM. We first exhibit the benchmark problems.

4.1   Benchmark Problems

Random. We generated binary random target networks with 50 variables, domains of size 10, and m binary constraints. We use a bias based on the language $\Gamma = \{=, \neq \leqslant, \geqslant, <, >\}$. For densities $m = 12$, we have launched our experiments.

Sudoku. The target network of the well-known Sudoku problem has 81 variables with domains of size 9 and 810 binary $\neq$ constraints on rows, columns and

---

[1] www.choco-solver.org.

squares. We use a bias of $19,440$ binary constraints taken from the language $\Gamma = \{=, \neq \leqslant, \geqslant, <, >\}$.

Latin Square. A Latin square is an $n \times n$ array filled with n different Latin letters, each occurring exactly once in each row and exactly once in each column. We have taken $n = 10$ and the target network is built with 900 binary $\neq$ constraints on rows and columns. We use a bias of $297,000$ constraints based on the language $\Gamma = \{=, \neq, \leqslant, \geqslant, <, >\}$.

Graceful Graphs. (prob053 in [12]) A labeling f of the n nodes of a graph with q edges is graceful if f assigns each node a unique label from $0, 1, \ldots, q$ and when each edge (x, y) is labeled with $|f(x) - f(y)|$, the edge labels are all different. The target network has node-variables $x_1, x_2, \ldots x_n$, each with domain $\{0, 1, \ldots, q\}$, and edge-variables $e_1, e_2, \ldots e_q$, with domain $\{1, 2, \ldots, q\}$. The constraints are: $x_i \neq x_j$ for all pairs of nodes, $e_i \neq e_j$ for all pairs of edges, and $e_k = |x_i - x_j|$ if edge e_k joins nodes i and j. The constraints of B were built from the language $\Gamma = \{\neq, =, \shortparallel_{xy}^z, \rtimes_{xy}^z\}$ where $\shortparallel_{xy}^z$ and $\rtimes_{xy}^z$ denote respectively the distance constraints $z = |x - y|$ and $z \neq |x - y|$. We used three instances that accept a graceful labeling: $GG(K_4 \times P_2)$, $GG(K_5 \times P_2)$, and $GG(K_4 \times P_3)$, whose number of variables is 24, 35, and 38 respectively, and bias size is 12,696, 40,460, and 52,022 respectively.

4.2 Results

In our first experiment, we work with Graceful graphs instances. Here, we compare the number of times we execute solve in τ ($\#solve_\tau$) and the number of times we execute solve of line 21 ($\#solve$). Our aim is to show that the line 21 allows our approach to avoid reaching a premature convergence and to learn all constraints. We report the results for $\tau = 0.1\,\text{ms}$, $1\,\text{ms}$, and $5\,\text{ms}$ with the parameters $\beta = 0.8$, time_bound $= 1\,\text{s}$, and the query size ℓ was initialized to $|X|$ (Table 1).

Table 1. $\#solve_\tau$ versus $\#solve$ on Graceful graphs instances

τ	Benchmark	$\#solve_\tau$	$\#solve$
0.1 ms	$GG(K_4 \times P_2)$	204	45
	$GG(K_5 \times P_2)$	301346	59
	$GG(K_4 \times P_3)$	289332	70
1 ms	$GG(K_4 \times P_2)$	2741	0
	$GG(K_5 \times P_2)$	66948	41
	$GG(K_4 \times P_3)$	78280	58
10 ms	$GG(K_4 \times P_2)$	1544	0
	$GG(K_5 \times P_2)$	5947	1
	$GG(K_4 \times P_3)$	6860	1

We observe that the smaller τ, the greater the number of times we execute the line 21 (#solve). The explanation is that at the end of the learning process, the more τ is small, the more is hard to solve the subproblem $C_L[Y] \cup B[Y]$ in τ. So, our algorithm spent time looping until reaching the time_bound. This leads to increase the number of times of solving in τ (#solve$_\tau$). Next, we execute the line 21 to generate a query. We deduce that thanks to line, P_a-QUACQ keeps learning constraints. If we have lunched this tests with T-QUACQ, we will get *premature convergence*.

In [1], it has been shown that T-QUACQ with the parameters $\beta = 0.8$, $\tau = 50$ ms lead to the best performance of T-QUACQ in terms of convergence. So, under this circumstance T-QUACQ and P_a-QUACQ will have the same performance. Thus, it is worth to compare T-QUACQ and P_a-QUACQ when the risk of reaching *premature convergence* in T-QUACQ is significant. Therefore to present clearly the efficient of our approach, we compare the results with QUACQ algorithm t. We choose τ to be small: 0.005 ms. Table 2 displays the performance of QUACQ and P_a-QUACQ. We report the total time $totT$, $MT(q)$ the maximum waiting time between two queries, #q the total number of asked queries, and #solve in P_a-QUACQ.

Table 2. P_a-QUACQ versus QUACQ, $\alpha = 0.8$, $\tau = 0.005$ ms

Benchmark ($|X|, |D|, |C|$)	Algorithm	$totT$ (sec)	$MT(q)$ (sec)	#q	#solve
rand-50-10-12 (50, 10, 12)	QUACQ	204	5.01	253	-
	P_a-QUACQ	17	2.22	138	8
Sudoku 9×9 (81, 9, 810)	QUACQ	2,820	1,355	9,053	-
	P_a-QUACQ	12	2.66	5769	2
Latin-Square (100, 10, 900)	QUACQ	7,200	1,234	12,204	-
	P_a-QUACQ	23	2.58	8,996	1

If we compare P_a-QUACQ to QUACQ, the main observation is that our approach reduces significantly the maximum waiting time between two queries.

Let us take sudoku problem. QUACQ takes more than 20 min to generate a query, where P_a-QUACQ generates all queries in a time less than 3 s. We notice also that the maximum waiting time between two queries in P_a-QUACQ is greater than the time_bound. The explanation is that P_a-QUACQ executes the line 21 which is not bounded by time cutoff.

The second observation, is the fact that P_a-QUACQ wins in the total time needed to learn the whole target network. If we take the latin Square instance, we observe that QUACQ needs more than 2 h instead of 23 s in P_a-QUACQ.

The last observation we can made, is that P_a-QUACQ is better than QUACQ in number of queries. Thanks to the use of time cutoff and the adjustment of the query size during the learning process.

5   Conclusion

We have proposed a new approach P_a-QUACQ, to make constraint acquisition more efficient by reducing the time of generating queries. Our approach has several advantages work. First, it always converges on the target constraint network. Second, the maximum waiting time between two queries is tolerable. Finally, the number of queries is small.

References

1. Addi, H.A., Bessiere, C., Ezzahir, R., Lazaar, N.: Time-bounded query generator for constraint acquisition. In: Integration of Constraint Programming, Artificial Intelligence, and Operations Research - 15th International Conference, CPAIOR 2018, Delft, The Netherlands, June 26-29, 2018, Proceedings. Lecture Notes in Computer Science, vol. 10848, pp. 1–17. Springer (2018)
2. Bessiere, C., Lazaar, N., Koriche, F., O'Sullivan, B.: Constraint acquisition. Artificial Intelligence (2017, in Press)
3. Bessiere, C., Daoudi, A., Hebrard, E., Katsirelos, G., Lazaar, N., Mechqrane, Y., Narodytska, N., Quimper, C.-G., Walsh, T.: New approaches to constraint acquisition. In: Data Mining and Constraint Programming - Foundations of a Cross-Disciplinary Approach. Lecture Notes in Computer Science, vol. 10101, pp. 51–76. Springer (2016)
4. Arcangioli, R., Bessiere, C., Lazaar, N.: Multiple constraint acquisition. In: Proceedings of the Twenty-Fifth International Joint Conference on Artificial Intelligence, IJCAI 2016, New York, NY, pp. 698–704 (2016)
5. Bessiere, C., Coletta, R., Hebrard, E., Katsirelos, G., Lazaar, N., Narodytska, N., Quimper, C.-G., Walsh, T.: Constraint acquisition via partial queries. In: Proceedings of the 23rd International Joint Conference on Artificial Intelligence, IJCAI 2013, Beijing, China, pp. 475–481 (2013)
6. Lallemand, C., Gronier, G.: Enhancing user experience during waiting time in HCI: contributions of cognitive psychology. In: Proceedings of the Designing Interactive Systems Conference, DIS 2012, New York, NY, USA, pp. 751–760. ACM (2012)
7. Beldiceanu, N., Simonis, H.: A model seeker: Extracting global constraint models from positive examples. In: Proceedings of the 18th International Conference on Principles and Practice of Constraint Programming, CP 2012. Lecture Notes in Computer Science, Québec City, QC, Canada, vol. 7514, pp. 141–157. Springer (2012)
8. Lallouet, A., Lopez, M., Martin, L., Vrain, C.: On learning constraint problems. In: Proceedings of the 22nd IEEE International Conference on Tools with Artificial Intelligence, ICTAI 2010, Arras, France, pp. 45–52 (2010)
9. Shchekotykhin, K.M., Friedrich, G.: Argumentation based constraint acquisition. In: Proceedings of the Ninth IEEE International Conference on Data Mining, ICDM 2009, Miami, FL, pp. 476–482 (2009)
10. Bessiere, C., Coletta, R., O'Sullivan, B., Paulin, M.: Query-driven constraint acquisition. In: Proceedings of the 20th International Joint Conference on Artificial Intelligence, IJCAI 2007, Hyderabad, India, pp. 50–55 (2007)
11. Freuder, E.C., Wallace, R.J.: Suggestion strategies for constraint-based matchmaker agents. Int. J. Artif. Intell. Tools **11**(1), 3–18 (2002)
12. Jefferson, C., Akgun, O.: CSPLib: a problem library for constraints (1999). http://www.csplib.org

FPSO-MPC Control of Artificial Pancreas

M. El Hachimi$^{(\boxtimes)}$, M. Tassine, A. Ballouk, and A. Baghdad

Laboratory of Electronics, Energy, Automatic and Data Processing (LEEA&TI),
FST- Mohammedia Hassan II University,
BP 146, 20650 Mohammedia, Morocco
hachimi@gmail.com

Abstract. Model Predictive Control (MPC) algorithm attracts the interest of researchers due to their flexibility and efficiency in solving control problems. However, the auto-tune of MPC parameters still a challenging task. This paper presents a new tuning of MPC using Fuzzy Particle Swarm Optimization (FPSO) implemented successfully in the Artificial Pancreas (AP). Our approach uses a fuzzy system to calculate the value of the inertia weight of PSO. Tree MPC parameters are adjusted using the FPSO. Experimental results illustrate the success of this method to improve the controller's performances compared to the previous ones.

Keywords: Model Predictive Control · Fuzzy Logic · Prediction Horizon · Particle Swarm Optimization · Artificial Pancreas

1 Introduction

The body become incapable of producing insulin, a hormone that facilitates glucose absorption from the bloodstream to many types of cell, and participates in the endocrine feedback loop that regulates the liver's release/removal of glucose into and from the blood stream, In this case we are speaking about Type 1 Diabetes Mellitus (T1DM) which is a metabolic autoimmune disease characterized by destruction of the pancreas' beta cells. The International Diabetes Federation (IDF), estimates that diabetes affects currently 415 million people and is set to escalate to 642 million by the year 2040 [1].

An individual with type 1 diabetes is confronted to long term risks associated with hyperglycemia (vascular problems, retina diseases), and short-term risk of hypoglycemia (drowsiness, coma). Individual must be diligent with frequent blood glucose meter (finger stick) tests, and insulin dosage adjustments (at mealtime as "correction" boluses). For tight control of blood glucose an individual must be diligent, and constantly managing their disease.

The Artificial Pancreas can overcome all this problems. This system will manage automatically insulin infusion in response to glucose changes, thus reduce patient burden. The Artificial Pancreas (AP) is composed of a subcutaneous glucose sensor (CGM), an insulin pump, and a control algorithm [2]. Figure 1 gives an overview of the Artificial Pancreas.

© Springer Nature Switzerland AG 2019
F. Khoukhi et al. (Eds.): AIT2S 2018, LNNS 66, pp. 257–272, 2019.
https://doi.org/10.1007/978-3-030-11914-0_28

Continuous Time Blood Glucose Measurement (CGM): CGM devices provides glucose readings in real time, it produces frequently-sampled data sets (e.g., every 5–10 min) allowing them to serve as AP-enabling technology, **CGM** can also display trends and blood glucose rate of change and is capable to alert the patient about upcoming hyper- or hypo-glycemia.

Pump Injection: Insulin pumps allow automatic insulin delivery. There are several technologies that can perform this task: an intra-venous route, subcutaneous insulin infusion **(SCII)** or intra-peritoneal insulin delivery. Continuous subcutaneous insulin infusion **(CSII)** uses a portable electromechanical pump to mimic non diabetic insulin delivery as it infuses at preselected rates normally a slow basal rate with patient-activated boosts at mealtime.

Control Algorithm is a software embedded in an external processor that receives information from the CGM and realizes a series of mathematical calculations. Based on these calculations, the controller send recommendation to the infusion pump to inject the computed amount of insulin. The system is subject to important disturbances, like meal, stress and exercises. This work is interested to the development of this part of AP.

The implementation of the controller in this system constitutes a challenged task because of some complex characteristics of the system (e.g. difficult dynamics, inter-action between process variables, the presence of disturbances, significant inter-individual variability and time delays due to the absorption of insulin from the subcutaneous level to the blood).

PID and MPC are the most used control algorithm in the development of Artificial Pancreas. Pinsker et al. realize a comparison between MPC and PID control, and indicate that MPC performed particularly well [3] in this system. Review of algorithm used in glucose control are provided by authors in [4].

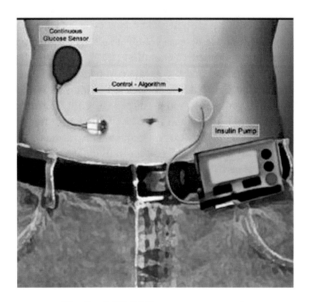

Fig. 1. Artificial Pancreas components

MPC becomes one of the main control strategies used in industries due to its intuitive control concept. Recently, MPC is implemented in many successful applications, including chemicals, aerospace, automotive and food processing applications [5].

MPC parameters such as weighting factors, control horizon Nu and prediction horizon Np play an essential role in the dynamic behavior of the closed-loop system. These parameters must be judiciously chosen to guarantee the stability and robustness of the controller. In general, their tuning is made in an empirical way, which does not lead to an optimal tuning. Therefore the use of an optimization algorithm to provide an appropriate values of the controller parameters is required.

The paper of Garriga [6] presents a review of the available tuning guidelines for MPC, from a theoretical and a practical view. Tran et al. [7] introduce a tuning of parameters using an optimal bandwidth that realize a trade-off between nominal performance and robustness. Davtyanetal. [8] proposed a method to optimally tune the parameters of an unconstrained MPC: prediction horizon, control horizon and control interval. Their aim was to test various performance objectives so as to select the most adequate criterion to solve the MPC tuning problem. Lee et al. [9] tuned MPC using a genetic algorithm (GA) technique to handle the desired multiple performance criteria in a multi-objective fuzzy decision-making framework.

This article introduces a new tuning of MPC based on Fuzzy Particle Swarm Optimization. The synthesized controller is implemented in Artificial Pancreas (AP). The simulation of the FPSO-MPC demonstrate the capability of the algorithm to avoid overshoot and to reject rapidly disturbances caused by meal.

The reminder of the paper is organized as follow: In Sect. 2 we present the modelization part. In Sect. 3 we introduce our controller. The tuning of MPC is presented in Sect. 4. Section 5 summarizes simulations and results of this work. Section 5 draws conclusions.

2 Modelisation

In order to develop a robust and safe controller, two models have been used. The first one is the model of Dalla Man et al. [10] used to design a simulation platform. It will be considered as a realistic virtual patient to evaluate the controller performances. It is a part of the framework of a platform of simulation approved by the Food and Drug Administration (FDA) [11]. The second one is the internal model of MPC used to predict the future evolution of glycemia within the prediction horizon. It is an ARX model identified from a priori information.

2.1   The Dalla Man et al. Model

The model of Dalla-Man et al. [10] is a realistic model representing human metabolic, it was obtained using complex experiments and it is composed from several compartments which interact among them to design accurate model of the glucose-insulin metabolism.

The model is a MISO (multiple input single output) system composed from two inputs (sugar consumption and insulin injection) and one output that represents the blood glucose. It was implemented in a virtual testing platform that was approved by the FDA [11]. The test in this platform is equivalent to animals test.

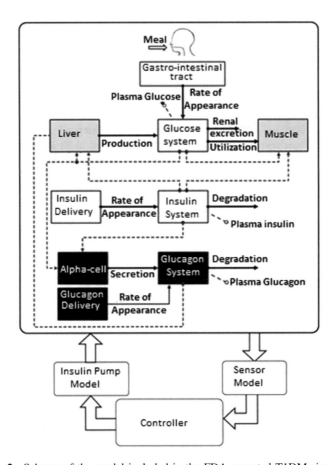

Fig. 2. Scheme of the model included in the FDA-accepted T1DM simulator

Figure 2 gives an overview of the glucose-insulin metabolism. The model structure is divided into three main compartments; insulin, glucose and meal. The meal and the insulin parts interact with the glucose part and influence the blood glucose evolution, relatively if the insulin quantity increases the blood glucose descent and if a meal is consumed the blood glucose rises.

2.2   Model for Control

Multiple control-relevant models have been developed to design the control algorithms for AP, but the predictions of these models still not perfect. considering the asymmetry in the glycemic scale because of the rapid effect of hypoglycemia that occurs in a very small range of glycemia, and the long-term effects of hyperglycemia that are seen in a large range. This is especially inherent because of the dangerous effect of excess insulin, and the incapacity to remove insulin from the subject after delivery [12].

We chose to use ARX model as a model of control because ARX models are simple with reduced computation time, and they are accurate enough to represent the actual insulin-glucose dynamics.

This model is identified from information available in the UVa/Padova simulator, the input is the variation of insulin considering the basal rate and the output is the variation of glucose level, Parameters of this model are chosen conservatively to minimize the probability of hypoglycemia events, on reverse of most models that minimize a prediction error. The model is personalized using a priori patients' characteristics in order to limit the conservatism due to large inter-subject variability.

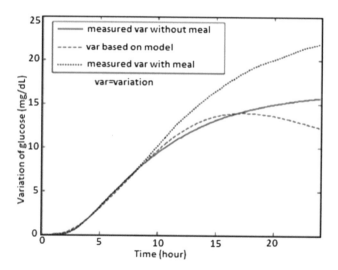

Fig. 3. Comparison of the ARX model to the real dynamics

2.3   Insulin-Glucose Transfer Function

The plant is a discrete time and linear time-invariant (LTI) system using a sample-period Ts = 5 min. The model is linearized around a steady-state that has been reached by applying the basal rate specific to the subject U_{BASAL} U/h, and lead to a steady-state output ys = 110 mg/dL. The input of the plant is the administered insulin bolus $U_{IN,i}$ [U] delivered per sample-period, and the output of the model is the subject's blood-glucose value $Y_{BG,i}$ [mg/dL].

The transfer function of the system is:

$$\frac{y(z^{-1})}{u(z^{-1})} = \frac{1800\,F.c}{U_{TDI}} \cdot \frac{z^{-3}}{(1 - p_1 Z^{-1})(1 - p_2 Z^{-1})^2} \tag{1}$$

With: $c = -60(1 - p_1)(1 - p_2)^2$, $F = 1.5$, $p_1 = 0.98$, $p_2 = 0.965$. $U_{TDI}[U]$ is the subject specific total daily insulin amount.

2.4   State-Space Model

The transformation from function transfer to state-space model leads to the following system:

$$x_{i+1} = Ax_i + B(u_i + d_i) \tag{2a}$$

$$y_i = Cx_i + n_i \tag{2b}$$

$$A := \begin{bmatrix} p_1 + 2p_2 & -p_1 p_2 - p_2^2 & p_1 p_2^2 \\ 1 & 0 & 0 \\ 0 & 1 & 0 \end{bmatrix} \in \mathbb{R}^{3 \times 3}$$

$$B := \frac{1800\,F.c}{u_{TDI}} \cdot [1\,0\,0]^T \in \mathbb{R}^3$$

$$C := [0\,0\,1] \in \mathbb{R}^{1 \times 3}$$

A, B and C are the state matrix describing model (2). The state is $x_i \in \mathbb{R}^3$. The inputs are the control input u_i that belongs to the set $\mathcal{U}_i \subset \mathbb{R}$ and the unmeasured disturbance d_i that belong to the set $\mathcal{D}_i \subset \mathbb{R}$. The measured noise n_i belong to the set $\mathcal{N}_i \subset \mathbb{R}$.

Figure 3 provides a comparison between glucose variation in batch direction without meal; glucose variation with three constant meals and estimated glucose variation based on the ARX model during 24 h. Result confirm that the proposed ARX model realizes an acceptable performance in the estimation of glucose variation.

2.5   State-Estimation

The Luenberger observer (a linear state-estimator) is used to estimate the initial state for each iteration. The state estimator is implemented as:

$$\hat{x}_{k+1} = A\hat{x}_k + BU'_k - L(Y'_k - \hat{Y}'_k) \tag{3}$$

$$\hat{Y}'_k = C\hat{x}_k \tag{4}$$

$$L = K^T \tag{5}$$

$$K = -(CPC^T + \hat{R})^{-1}CPA^T \tag{6}$$

$\hat{x}_k$ represents the estimated states of X_k and $\hat{Y}'_k$ represent the estimated BG Y_k, L is the observer gain. P satisfies the discrete algebraic Riccati equation:

$$P = APA^T + \hat{Q} - APC^T(CPC^T + \hat{R})^{-1}CPA^T \tag{7}$$

Where $\hat{Q} = 1$ and $\hat{R} = 1000$ are positive definite design parameters.

3   Controller Design

3.1   Concept of MPC

In this work we use MPC to control blood glucose because it can predict future evolution of system, compensate time delay existing in glucose-insulin system, manage patient's variability and introduce hard constraints on insulin delivery [13].

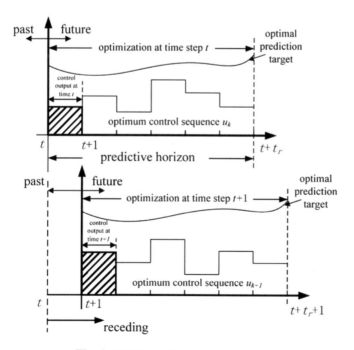

Fig. 4. Model predictive control process

MPC is an intelligent control technique developed in the 1970 s. It is a part of model based methods that uses a model to predict the future state of the system $\mathbf{y_k}$ along the prediction horizon Np then minimizes an objective function $\mathbf{J(u)}$ (weighted sum of tracking error of predictive outputs y_k to set points y_{ref}, penalized with the input weight $\mathbf{R}$ and the output weight $\mathbf{Q}$) to compute the optimal control $\mathbf{u_k}$ along the control horizon $\mathbf{Nu}$, then implements only the first element of $\mathbf{uk}$ [14]. The concept of MPC is shown in Fig. 4.

3.2   Cost Function

The formulation of the optimization problem is:

$$\{u_0^*, \ldots, u_{Nu-1}^*\} := arg \min_{\{u_0, \ldots, uN_{u-1}\}} J(x_i, \{u_0, \ldots, u_{N_{u-1}}\}) \tag{8}$$

With cost function:

$$J(.) := \sum_{k=1}^{Np(k)} Q(k)(y_k - y_{ref})^2 + \sum_{k=0}^{N_{u-1}} R(k)u_k^2 \tag{9a}$$

Subject to

$$x_0 := x_i \tag{9b}$$

$$x_{k+1} := Ax_k + Bu_k \qquad \forall k \in \mathbb{Z}_0^{N_p-1} \tag{9c}$$

$$y_k := Cx_k \qquad \forall k \in \mathbb{Z}_0^{N_p} \tag{9d}$$

$$0 \leq u_k + u_{BASAL}(i+k) \leq u_{max} \qquad \forall k \in \mathbb{Z}_0^{N_{u-1}} \tag{9e}$$

Where: $\mathbb{Z}_+$ is the set of positive integers. $\mathbb{Z}_a^b$ is the set of consecutive integers $\{a, \ldots, b\}$. $N_p \in \mathbb{Z}_+$ is the prediction horizon. $N_u \in \mathbb{Z}_1^{Np}$ is the control horizon. u, x, y are respectively the predicted input u, the state x and the glucose output y. $R \in \mathbb{R}_{>0}$ is the penalization of the control inputs, $Q \in \mathbb{R}_{>0}$ is the penalization of the system output. Equations (9a)–(9d) predict the future state of the system. Equation (9e) represent input constraint.

The optimization function of Matlab used in this work is fmincon function to solve constrained problem by applying the following hard constraint (u):

$$-u_{BASAL} \leq u_k \leq 72\,u/h \tag{10}$$

3.3   Role of MPC Parameters

The horizon of prediction **Np** specifies the length of time steps that the predictive controller has to drive the state to its reference values. Wojsznis et al. [15] affirm that **Np** has to be adjusted large enough, so that further increment doesn't affect control performance. In fact, large values of **Np** generally improve the stability and robustness, but increase the computational time of the controller as well. There are some cases reported in the literature which confirm that an increase of Np could deteriorate the controller robustness.

Q provides additional degrees of freedom to solve the tuning problem. Small value of **Q** yield to a smooth output and large value yield to a faster output tracking. A correct setting of R helps prevent both aggressive control and an oscillatory behavior of the controlled outputs. Moreover, increasing **R** can cause a sluggish control performance, and may produce unacceptable effects on the closed-loop system. Figure 5 presents a comparison between different control settings of Q: R ratio of MPC in response to a typical meal disturbance of 60-g CHO.

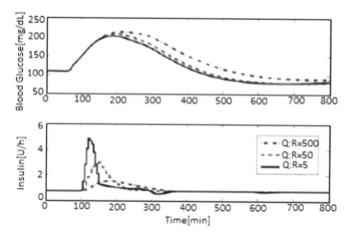

Fig. 5. Comparison of different control settings of Q:R

Seen the significantly impacts of MPC parameters on the closed loop performance, a good tuning of this parameters will improve the set-point tracking and disturbance rejection, thus a FPSO tuning is proposed in the next section.

4   Tuning of MPC Parameters

4.1   Particle Swarm Optimization

Particle Swarm Optimisation (PSO) is a stochastic optimization technique based on the social behavior of swarms of flocking animals. It was originally proposed by Kennedy and Eberhart in 1995. The PSO theory is based on the observation of collective behavior of some animals, such as bees, flocks of birds and ants. Particles adjust their trajectory towards its own best location and towards the best particle of the swarm at each iteration of the process. The swarm direction of a particle is defined by the set of particles neighboring the particle and its history experience [16].

The velocity and the new position of every particle used in this work are respectively:

$$V_{ij}(k+1) = wV_{ij}(k) + C_1r_1\big(P_{il}(k) - x_{ij}(k)\big) + C_2r_2\big(P_{ig}(k) - x_{ij}(k)\big) \qquad (11)$$

$$x_{ij}(k+1) = x_{ij}(k) + V_{ij}(k+1) \qquad (12)$$

$P_{ig}(k)$ and $P_{il}(k)$: are respectively the global and the personal best position.

c_2 and c_1: are the learning factors that determine the impact of the global best Gbest and the personal best Pbesti, respectively. If $c_1 < c_2$, the particle has the tendency to converge to the best position found by the population Gbest rather than the best position found by itself Pbesti, and vice versa. Many studies use a setting with $c_1 = c_2 = 2$ [17].

w: is the inertia weight that balances local exploitation and global exploration abilities of the swarm. A small inertia weight makes the convergence fast; however it sometimes leads to a local optimal, and vice versa.

4.2 Fuzzy PSO

When we use PSO with a fix inertia weight, we don't have a feedback about the distance between the fitness of the particles and the optimal real values, furthermore PSO presents some drawback in the search of local optimum. Therefore, we use Fuzzy Logic to update the inertia weight of PSO. Fuzzy logic can represent experience and expert knowledge in some linguistic rules to update inertia weight. Figure 6 shows the architecture of fuzzy controller.

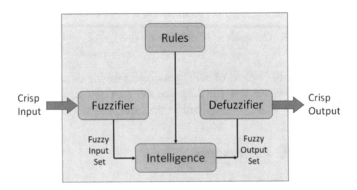

Fig. 6. Architecture of fuzzy controller

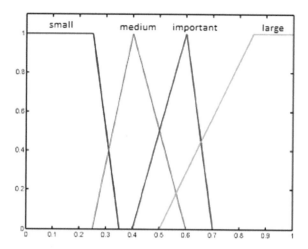

Fig. 7. Membership functions of the $NFCB_i^K$ input

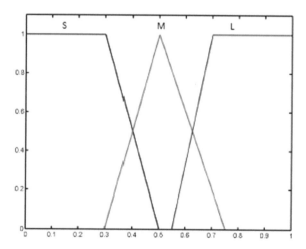

Fig. 8. Membership functions of the w_i input

We designed two-inputs and one-output in our Fuzzy Logic System. The two-inputs variables mentioned here are the normalized fitness of the current best position ($NFCB$) and the inertia weight ω_i; the unique output is the inertia weight variation $\Delta\omega_i$.

Equation 13 represents the normalized fitness of the current best position ($NFCB$).

$$NFCB_i^K = \frac{fitness\left(pbest_i^k - F_{KN}\right)}{fitness\left(pbest_i^1 - F_{KN}\right)} \tag{13}$$

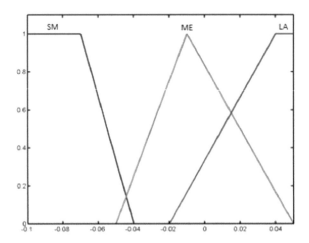

Fig. 9. Membership functions of the Δw_i output

fitness $pbest_i^k$ is the current best previous position's fitness. F_{KN} is the known real optimal solution value. **fitness** $pbest_i^1$ is the fitness in the first iteration.

Figures 7, 8 and 9 represent membership functions of the inputs and the output.

Three membership functions, 'Large' 'Medium' and 'Small' are used for the w_i input, see Fig. 8. Figure 9 illustrates the variation of the inertia weight $\Delta \omega_i$.

The inference part defines rules describing the system. We define twelve rules using the min-max Mamdani method. Rules are illustrated in Table 1. The defuzzification part computes the output using the centroid of sets method with the following equation.

$$\Delta \mathbf{w_i} = \frac{\sum_{i=1}^{12} \mathbf{w^i} \Delta \mathbf{w_i}}{\sum_{i=1}^{12} \mathbf{w^i}} \tag{14}$$

Table 1. Fuzzy rules of FPSO

w_i	$NFCB_i^K$			
	Small	Medium	Important	Large
S	LA	LA	LA	LA
M	SM	ME	ME	ME
L	SM	SM	SM	SM

4.3   Flowchart of the Proposed Controller

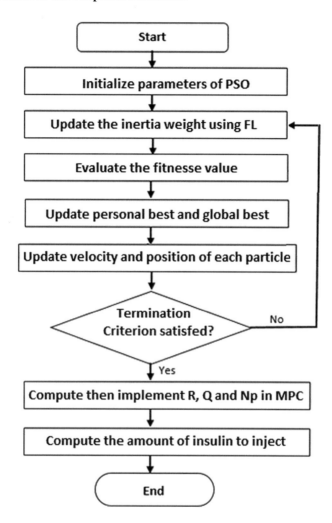

5   Results and Discussion

5.1   Metabolic Simulator System

The presented controller is implemented under Matlab in a simulation platform developed by the University of Padova, the simulator was approved by the U.S. Food and Drug Administration (FDA) for verification of control algorithms for AP before doing clinical trials on T1DM subjects [11]. Figure 8 gives an overview of blocks of the controller implementation.

5.2   Protocol of Simulation

The proposed controller is compared to the non-adapted MPC, which is characterized by a constant penalization with R = 100, Q = 0.8, Np = 9, Nu = 5 and yref = 110 [18]. 30 virtual patients were under 30 h of simulation. Controlled by two algorithms the FPSO-based MPC controller and the non-adapted MPC. During this period, they took three meals: (1) dinner (45 gCHO) at 7:30 PM, (2) breakfast (30 g carbohydrate gCHO) at 7:00 AM and (3) lunch (55 gCHO) at 12:00 PM.

5.3   In Silico Results

The evolution of the mean blood glucose and insulin delivery during simulation are given in Fig. 10. The evolution of glycemia shows that after the meal intake the glycemic reaches **240** mg/dl in the case of standard MPC, however for the proposed MPC it doesn't exceed **220** mg/dl. The mean glucose profile obtained with FPSO-MPC outside post prandial periods is always characterized by values below the upper limit (140 mg/dl) of the tight target range. In fact, after each meal the mean glucose returns below 140 mg/dl before the end of the post prandial period. On the contrary of the non-adapted MPC case where the mean glucose profile is generally higher after meals and exceeds the upper limit of the tight target range even after the end of the postprandial period.

5.4   Time in Euglycemic Target Range [70–180] (mg/dl)

FPSO-MPC achieves a percentage of overall time in this target range from **79% to 86%** counting the post prandial period. The most critical time window is the post-dinner one especially (randomly perturbed insulin sensitivity) where percent time in target is **71%.** On reverse, for the non-adapted MPC, the overall time in target ranges is from **74% to 80%,** the most critical time window being the post-dinner one (time in target: **60%**). If we focus the attention on the night time, it is seen that all the both strategies achieve nearly optimal performances, as the percent time in target ranges is from **88% to 90%** across the both methods.

5.5   Hyper- and Hypo-Glycemia

The period spent above 150 mg/l in the case of standard MPC is 3 h whereas in the case of FPSO-MPC, it is limited to 2 h. We remark also that the amount of insulin injected by FPSO-MPC is less than the non-adapted MPC case because the command is accelerated and the need of insulin is reduced. Furthermore the post-prandial peak of glycemia is dropped compared to the classical controller. It is seen that the tuning controller strategies succeed in avoiding hyper- and hypo-glycemic episodes, above 250 and below 70 mg/dl, respectively.

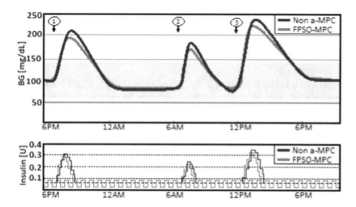

Fig. 10. Simulation progress [h]: Mean glucose (A) and Insulin delivery (B) for 30 patients controlled by FPSO-MPC and non-adapted MPC

5.6   General Discussion

Further analysis shows that after perturbation caused by a meal, the FPSO increases the output weight and decreases the input weight to accelerate the reference tracking, then a large amounts of insulin is automatically injected as early as possible to restore blood sugar to their normal level and reduce the glucose peak and time spent in hyperglycemia. The prediction horizon increases under the same conditions to make the control more stable and to better predict the evolution of the blood glucose with a small increase in computation time due to the increment of MPC iterations.

By comparing the two controller's performances, it is observed that the non-adaptive predictive controller has several limitations in the control of blood glucose: delay in rejection of disturbance, some hypoglycemia due to insulin over-delivery and hyperglycemia. One reverse, the proposed controller brings several advantages to the AP development. It accelerates control when a change in blood glucose level happens and makes it very conservative when the system starts to recover. The FPSO-based MPC overcomes limitations of the non-adapted controller and lead to a rapid reference tracking with a reduction of overshoot and a complete avoidance of hypoglycemia. The proposed method doesn't react when the glycemia is in the euglycemic range thus optimizes the functioning of the pump and rejects different disturbance, with this method small variations don't affect the control.

6   Conclusion

This work presents a new tuning of MPC based on the fuzzy particle swarm optimization to control the delivery of insulin with an Artificial Pancreas. The value of the inertia weight of PSO is determined by a set of fuzzy rules then MPC parameters are computed to ameliorate performance of the controller. This method is not sensitive to the initial choice of the parameters and a deep knowledge of the MPC concept is not necessary. Simulation shows the success of such method to overcome control difficulties by reducing overshoot of glycemia and avoiding hypoglycemia.

References

1. International Diabetes Federation. IDF Diabetes Atlas, 7th edn. cited 2017 April. http://www.diabetesatlas.org
2. Doyle, F.J., Huyett, L.M., Lee, J.B., Zisser, H.C., Dassau, E.: Closed-loop artificial pancreas systems: engineering the algorithms. Diabetes Care **37**, 1191–1197 (2014). https://doi.org/10.2337/dc13-2108
3. Pinsker, J.E., Lee, J.B., Dassau, E., Seborg, D.E., Bradley, P.K., Gondhalekar, R., Bevier, W.C., Huyett, L., Zisser, H.C., Doyle III, F.J.: Randomized crossover comparison of personalized MPC and PID control algorithms for the artificial pancreas. Diabetes Care **39**, 1135–1142 (2016)
4. El Hachimi, M., Ballouk, A., Lebbar, H.: Control algorithm of artificial pancreas systems - a review. Int. J. Multidisciplinary Sci. Issue 4(p1–p9) (2016)
5. Qin, S.J., Badgwell, T.A.: A survey of industrial model predictive control technology. Control Eng. Pract. **11**, 733–764 (2003)
6. Garriga, J.L., Soroush, M.: Model predictive control tuning methods: a review. Ind. Eng. Chem. Res. **49**(8), 3505–3515 (2010)
7. Tran, Q.N., Özkan, L., Backx, A.C.P.M.: MPC tuning based on impact of modelling uncertainty on closed-loop performance. In: AIChE Annual Meeting, 186e. Pittsburgh, PA (2012)
8. Davtyan, A., Hoffmann, S., Scheuring, R.: Optimization of model predictive control by means of sequential parameter optimization. In: IEEE Symposium on Computational Intelligence in Control and Automation, pp. 11–16 (2011)
9. van der Lee, J.H., Svrcek, W.Y., Young, B.R.: A tuning algorithm for model predictive controllers based on genetic algorithms and fuzzy decision making. ISA Trans. **47**(1), 53–59 (2008)
10. Dalla Man, C., Rizza, R.A., Cobelli, C.: Meal simulation model of the glucose-insulin system. IEEE Trans. Biomed. Eng. **54**, 1740–1749 (2007)
11. Kovatchev, B.P., Breton, M., Dalla Man, C., Cobelli, C.: In silico preclinical trials: a proof of concept in closed-loop control of type 1 diabetes. J. Diabetes Sci. Technol. **3**, 44–55 (2009)
12. Lee, J., Gondhalekar, R., Dassau, E., Doyle III, F.J.: Shaping the MPC cost function for superior automated glucose control. Ind. Eng. Chem. Res. **49**(7), 779–784 (2016)
13. Hovorka, R.: Continuous glucose monitoring and closed-loop systems. Diabet. Med. **23**(1), 1–12 (2006)
14. Forlenza, G.P., Deshpande, S., Ly, T.T., Howsmon, D.P., Cameron, F., Baysal, N., Mauritzen, E., Marcal, T., Towers, L., Bequette, B.W., Huyett, L.M., Pinsker, J.E., Gondhalekar, R., Doyle III, F.J., Maahs, D.M., Buckingham, B.A., Dassau, E.: Application of zone model predictive control artificial pancreas during extended use of infusion set and sensor: a randomized crossover-controlled home-use trial. Diabetes Care dc170500 (2017)
15. Wojsznis, W., Gudaz, J., Blevins, T., Mehta, A.: Practical approach to tuning MPC. ISA Trans. **42**(1), 149–162 (2003)
16. Kennedy, J., Eberhart, R.: Particle swarm optimization. In: Proceeding of the IEEE International Conference on Neural Networks, pp. 1942–1948 (1995)
17. Niknam, T.: A new fuzzy adaptive hybrid particle swarm optimization algorithm for non-linear, non-smooth and non-convex economic dispatch problem. Applied Energy (2009). https://doi.org/10.1016/j.apenergy.2009.05.016
18. Gondhalekar, R., Dassau, E., Doyle III, F.J.: Tackling problem nonlinearities & delays via asymmetric, state-dependent objective costs in MPC of an artificial pancreas. ScienceDirect IFAC-PapersOnLine **48**(23), 154–159 (2015)

Lean in Information Technology: Produce the Human Before the Software

Soukayna Belkadi[✉], Ilias Cherti, and Mohamed Bahaj

Department of Mathematics and Computer Science, Faculty of Sciences
and Technology, Hassan 1st University, Settat, Morocco
soukayna.belkadi@gmail.com

Abstract. Governments and private companies are under increasing pressure to improve their efficiency in delivering more and better services to citizens. Faced with the new industrial context characterized by the opening of markets, the liberalization of trade and the emergence of information technologies (IT); this companies are forced to be more flexible to survive. As a result, IT activities are becoming increasingly important in the operation of government business. Actually the Information systems have become more complex, especially with the introduction of software packages based on fully technical platforms, this complexity and duplication of business information can lead to additional costs in terms of development and maintenance. In this context that the combination between the management of information system and lean principles namely in piloting software development projects could bring more and more fruitful results and will seeks incremental waste reduction and value enhancement.

We will try in this paper to show how Lean, from the world of manufacturing, applies to the field of information systems security can improve the performance and the security of the information system. This paper will be broken into four sections: Sect. 1 discuss the evolution of lean from the world of manufacturing to the IT functions, the second section discuss the history and the challenges of the information system, the third section presents the IT governance tools and the limits of their application and the last section recommends a methodology to improve the IT maturity of an organization by discussing how Lean can drive the success of IT environments of organizations. It ends by presenting the challenges of the application of lean thinking to the IT functions.

Keywords: Lean thinking · Information system · IT governance ·
Management of Information System (MIS)

1 Introduction

Based on the Japanese economic model, Lean Management has been successfully applied by companies around the world, to ensure its continuous improvement, mainly in manufacturing functions. Recently, the digital transformation has been accelerating for 30 years and profoundly modifying our lifestyles and organizations. Companies are faced with having powerful and robust information systems.

Moreover, the information system is experiencing a real change with the arrival of New Information and Communication Technologies, This situation, coupled with the

© Springer Nature Switzerland AG 2019
F. Khoukhi et al. (Eds.): AIT2S 2018, LNNS 66, pp. 273–283, 2019.
https://doi.org/10.1007/978-3-030-11914-0_29

globalization of economies, creates a turbulent economic environment around companies. Thereby several parameters define the level of a company in the market, not only its productivity but also the robustness and security of its information systems.

It is in this context that the interest to investigate a wider application of Lean Management especially in service functions increased.

The Lean Approach ensures in particular the satisfaction of the customer through the elimination of any type of waste, this approach is often absent from the operational optimization of the IS Directions. However, it is not clear how Lean Management can be applied to IT organizations. Therefore, the goal of this paper is to translate the concept of Lean used in production functions to the field of service functions namely to the information system. We recommend adjusting Lean to the nature of IT management. Attention will be given to the amelioration of the productivity and the security of information system by using Lean principles.

2 The Evolution of the Lean Concept

Coming from Toyota's total quality method, like many others (Six Sigma, …), the Lean method is not revolutionary, but its principles are currently being updated wherever rationalization is needed, especially in computer science. In the Lean method, two principles drive the creation of value:

- The value must be estimated from the point of view of the customer;
- The value must "flow without interruption" along the design-production-distribution chain.

The origins of the Lean thinking go back to the Japanese manufacturers. (Shingo, 1981, Ohno, 1988, Monden, 1993). After 1990, the orientation of the workshop gradually widened, characterized by the design by many detractors. This evolution has focused on quality (early 1990s), cost and delivery (late 1990s), customer value from 2000. Toyota's philosophies have been shaped by the personalities, ethics and abilities of its creators in the Toyota family. Lean's principles are firmly grounded in scientific wisdom and methods (Bicheno and Holweg 2009) [3].

Moreover, the term "lean" has been increasingly popularized in the seminal book "The Machine Who Changed the World" (Womack, Jones and Roos, 1990), which emphasized the significant performance gap between Japanese and Western automotive industries. He described the key elements for a better industrial performance that is part of the Lean concept, as Japanese business methods used less than everything: human effort, investment, facilities, inventories and time in manufacturing, Product development, parts supply and customer relations.

In addition, the Lean concept is based on the concept of more productive and more economical management of processes and therefore of the company. It aims to satisfy customers by reducing the cost and investment of production, making the most of the resources, reducing the stock and the production cycle. Lean concept allows companies to reduce costs continuously, increase quality and shorten the supply cycle in order to meet customer expectations to the maximum and adapt to environmental conditions by eliminating all types of waste by using a lot of tools such as Kanban which is Material

Flow Control mechanism (MFC) to deliver the right quantity of parts at right time, Single Minute Exchange of Dies (SMED)/One-Touch exchange of Die (OTED) to reduce the changeover time by converting possible internal setting time (Carry out during machine stoppage) to external time (performed while the equipment is running, and the Value stream Mapping (VSM) which provides a clear and simple vision of the process [4].

Several books were inspired by the lean concept used in the industrial sector, to apply it to the information system. The beginning of Lean application in IT activities first appeared in Mary and Tom Poppendieck's book "Lean Software Development" published in 2003; they adapted manufacturing methods to software development throughout seven key principles of lean:

(1) Eliminate waste;
(2) Build integrity;
(3) See the whole;
(4) Decide as late as possible;
(5) Deliver as fast as possible;
(6) amplify learning;
(7) Empower the team.

It is in this context that the application of lean principles to the development of information system will help to control and secure the information system while eliminating any kind of waste that creates no added value. The role of the information system in the enterprise and the benefits of applying lean in information systems will be described in the next section.

3 History and Challenges of the Management of Information Systems

3.1 Evolution of Information Systems

The concept of management of information system appeared in the 1960s in the United States (in management department of Minnesota University). It was, then, soon adopted by many management academic centers as a modern scientific attitude. It was primarily designed for managing organizational applications. Hence the name "Management Information System" which is composed of three concepts: "management", "information" and "system". From the beginning, the main goal of MIS was to assist decision-making by presenting the scientific solutions and techniques required for the managers and decision-makers of the company to design, implement and manage automatic information systems in companies.

Indeed, the information system management is directly related to the concepts of decision support. The technical infrastructure of the systems, as the foundation of information management within the organization and the applications as business tools, quickly make it a strategic issue for companies [9].

3.2   Issues of the Information System

To meet the IT needs of a structure, the coherence and agility of the information system is primary to effectively integrate new requirements. From another point of view, it must at the same time integrate new technologies and the possibilities of data analysis, information processing and above all ease of implementation to support the processes.

Continuity of service after a disaster becomes a major issue also motivated by regulatory standards for the good health of a company. The Information Systems Department has the obligation to ensure that the technical standards and human procedures are respected during a disaster but also that the organizational management allows solving the problems quickly and efficiently. It is in this context that the security of information systems is a key issue for IT systems because the vulnerability of systems and human manipulation are essential factors that the Director of Information Systems (DSI) analyzes in depth, by asking about: the threats every company needs to pay attention to? What is the maturity of the processes put in place within an organization aimed at guaranteeing Confidentiality, Availability and Integrity of data?

The control of information systems is also one of the most important issues of the company. While computerization of enterprises could originally have been seen as a simple problem of automation of administrative tasks, the efficient and effective use of technologies has now become strategic and it concerns all organizations, whatever their size and area of activity [2].

The problem that companies are currently facing with their information system is that it has a growing number of applications running on different systems for more and more diverse users, which they use from increasingly diverse hardware platforms. In addition to these applications, the company often has a multitude of databases working with different technologies, several repositories on heterogeneous platforms. This heterogeneity is highlighted when the company concerned wishes to develop new applications by reusing as much as possible the existing components within the IS [1].

It is in this context that companies have moved from corporate governance to IT Governance, the goal is to build the decision-making system, focus alignment of business processes and setting up permanent communication. Referring to the most well-known governance tools that we will define in the next paragraph.

4   IT Governance Tools

4.1   Control Objectives for Information and Technology (COBIT)

This method was designed by the Information Systems Audit and Control Association (ISACA) about ten years ago. Which provide several tools allowing managers to bring control needs closer to technical solutions and risk management. It seeks to frame the entire informational process of the company from the creation of information to its destruction to ensure accurate quality monitoring.

The originality of CobiT is to create a link between stakeholders and CIO, which often requires a small cultural revolution for both actors of the Direction of information system in their ivory tower for trades and general management who would ignore superbly the strategic character of the IS. The key point behind this approach is the establishment of constructive dialogues at all levels of the organization, between stakeholders and CIOs [12].

The CobiT repository, with its 34 generic processes, is a proposal that can be revised to be adapted to the organization's own cartography. In the same way, CobiT can easily be linked to other market references (ISO 27001, ITIL for the Information Technology Infrastructure Library or CMMI for Capability Maturity Model Integration) by building a reference framework that meets all the requirements. This is all the more true since the processes of CobiT are sometimes global and are often interpreted as "macros" of more specialized references. CobiT is therefore a unifying framework.

4.2   Information Technology Infrastructure Library (ITIL)

Information Technology Infrastructure Library (ITIL) was created within the UK Public Trade Office, it offers a structured collection of best practices for the management of the ITIL information system, and a structured, process-oriented development framework focused on the customer. The customer is indeed the founding point and the heart of the approach, using ITIL allows companies to increase user and customer satisfaction with IT services and to improve service availability.

ITIL is the most widely adopted tool for ITSM. This practice can be adapted for use in all commercial and organizational environments. ITIL's Value Proposition Focuses on IT service provider (internal or external provider) including the objectives and priorities of a customer, and the role that IT departments in achieving these goals [10].

To sum up, ITIL improves the functions in all the entities of the company. Customers will be delighted with the improved quality of IT services through the execution of consistent and repeatable processes.

4.3   Capability Maturity Model Integration (CMMI)

Capability Maturity Model Integration (CMMI) is a model for evaluating the design process of software and allows the company to classify its practices in five levels of maturity: delay, quality, cost and reliability, its purpose is to measure the capacity of projects to be completed correctly in terms of deadlines, functionalities and budget.

CMMI helps to identify and achieve predefined business goals in the business by answering the question "how do we know?", it helps also to develop better products, get closer to the customer and meet their needs. It is composed of a set of "process areas" that must be appropriate to the company's policy. CMMI does not define how society should behave but defines what behaviors need to be defined. In this way, CMMI is a "behavioral model" and a "process model".

These governance tools are complementary; indeed COBIT's high-level control objectives can be achieved through the implementation of ITIL best practices and can be measured and evaluated by the CMMI model.

To meet the need of developing better products and get closer to the costumer, the management of information systems was defined in several ways by several authors that we will discuss below:

MIS is a system which receives data from different units and produces information and provides managers in all levels with relative, just-in-time, precise and uniform information for decision-making (Safarzade and Mansoori, 2009).

MIS is a manual or computer system which improves every organization's just-in-time use, management and processing of data and information (Feizi, 2005).

MIS involves official methods of providing precise and just-in-time information to facilitate managers' decision making processes while planning, controlling and making effective and optimal decisions in the organization (Momeni, 1993).

As can be seen, all definitions define MIS as a "system" with inputs and outputs whose purpose is to provide the right information at the right time, this principle is the main objective of the lean concept which aims in particular to get closer to the customer to provide the right product at the right time with the right price by eliminating as much as possible non-value-added tasks. This concept could therefore be associated with the information system where the product provided is the information that must be useful for its users.

5 Lean Concept Applied to the Information System

5.1 Mudas of the Information System

Lean Information Technology is the application of Lean manufacturing and Lean services principles to the development and management of information technology products and services. It designs the information system that could provide right information to the right people, at the right time and first times, it aims also, the elimination of waste that adds no value to a product or service by using particular principles and methods.

This method is complementary to other project management techniques like Scrum that allows teams to focus on delivering product and improved communication it has been designed for collocated software development and it focuses on project management institutions where it is difficult to plan ahead [8], unlike the Lean that was initially applied to the production systems and actually it can be transported to the information system; it allows companies to achieve permanent improvement and facilitates the work of teams, eliminate non-values added activities and reduce complexity of systems.

In general, when we consider developing a system by Lean Management Information system, we normally refer to all the tools of lean used in production system such as VSM, PCA, SMED and specially the Japanese 5S: Structurize, Systemize, Standardize, Self-Discipline, and Sanitize. These lean tools can help companies to minimize the sources of waste (Muda) in the information system defined in Table 1.

These sources of waste can be improved by using Lean practices, namely Failure Modes, Effects and Criticality Analysis (FMECA) to avoid failures due to unstable processes. Optimization packages to put into production (decoupling applications) can also help to minimize the waiting time in applications, implementation of processes and tools for skills management, implementation of capitalization tools and processes, the creation of sharing times between teams including the costumer and the implementation

Table 1. Sources of waste in the information system (MUDA)

Mudas	Examples
Defaults	• Repetitive activity • Unstable processes • Low level of customer relationship management • No compliance with cost/quality/time commitments
Waiting time	• Poor communication between the IT and the production Departments that causes • A delay between service and system development processes
Overproduction	• Deliveries of unused features • Bad definition of customer needs which generates a production earlier • Faster or in greater than the quantity expressed by the customer (Level of service too high (24 h/24 h–7days/7days for an application used on working days)
The unusual use of knowledge	• Low capitalization of knowledge • Non-reuse of components (codes…) • Lack of tools and skills management process
Activities without added value	• Activities that do not add value to customers, such as presenting technical dashboards to managers
Unnecessary movement	• Emergency response to recurring problems • Frequent change of environments (technical bases, framework, tools…) • Multi staffing: too many topics managed in parallel, no prioritization • Multiple interventions on incidents without resolution

of collaborative tools (wiki, collaborative workspace, search engines…) can help capitalize knowledge and process review by identifying and eliminating non-value-added tasks (VSM, MIFA).

5.2   Lean Applied to the IT Governance Tools

The adapted information system is based on the agile development approach, provide information in the field, especially to middle management to ensure good coordination between the decisions that emanate from the top management and their application in the market, and not just give reporting to the central Direction. The goal is to bring the decision and the decision maker closer to the market.

Therefore, agile development approaches can rely on Lean Management. These approaches are based on cycles of iterations with, for example, frequent deliveries (customer feedback), reusable automated tests, incremental designs, collaborative work, etc. Agile design requires companies to have flexible structures, with high technical quality, as opposed to large monolithic projects.

As we mentioned in the previous paragraph, ITIL, used in the agile method as a conceptual framework that promotes foreseen tight interactions throughout the development cycle. A series of books in the reference [5, 6] recommended practices on a broad range of IT management topics.

However, ITIL has some limits and is still reactive in nature, as a response to one or more incidents. It does not define the time consumption and does not provide insight into gaining efficiencies, nor does it address the leadership issues of organizational change. Furthermore, the implementation of an ITIL initiative can be difficult since it is a top down approach.

The CMMI also, unlike Lean IT, doesn't directly address sources of waste such as a lack of alignment between business units and the IT function or unnecessary architectural complexity within a software application, increased cost and overheads energy, waiting slow application. Then, using the Value Stream in the IT provides some services by the IT function to the organization that can be used by costumer, suppliers and employees, it analyses services into their component process steps.

It is in this context that the application of Lean management in a company can be associated with the development of information systems; indeed incremental developments also follow the iterative development of quality and processes.

Lean helps to control its software delivery times, to ultimately result in a streaming delivery. It can also help an IT support department to significantly reduce the time to correct incidents, by identifying and eliminating major waste in the chain of resolution and also by setting up trainings between team members to raise in competence all the collaborators.

Lean thinking uses a lot of tools to organize work namely:

- Kanban: First invented in the automotive sector, it is also possible to adapt Kanban to software creation. It helps to start where the team is in its maturity of knowledge of processes. It doesn't need to start a cultural revolution right away, as Scrum demands; Companies can keep the same people and the same roles. It can also be a solution for teams who cannot switch to Scrum for corporate cultural reasons but still want to be more agile.
- Collective visualization: One of the keys to Lean is the development of person's skills through job training or problem solving. Lean information management attaches great importance to the collective visualization of the (performance, problems…) and on its application to the computer science, where everything tends to happen in the computer and networks.

The lean principle in IT uses the principle of producing the human before producing software. This combination between lean and information system creates two value one for the customer by facilitating access and navigation on the IS and another value for employees by facilitating the processing and access to information, it allows to devote time to activities that really create value [11].

- Key Performance Indicator (KPI): It is generally difficult to measure the value for an employee working on an information system, the customer for example when he is not satisfied with the service offered by the IS, he sends complaints, comments on the internet…, which is not the case for an employee. Computer scientists are struggling to measure the benefits provided by the various functionalities of the developed IS in a company that has a poor knowledge of its operational performance, which can lead to provide useless information; employees are forced to obey the instructions issued by the enterprise system, they are then transformed not an active agent expert, but simple "clickers" obeying instructions. In this case the computer team does not bring added value to the IS [11].

- Value stream mapping and HEIJUNKA: help to know all source of waste and eradicate non value adding processing and poor costumer services, ovoid the over production and application changes and also leveling the customer demand by using the HEIJUNKA In order to overcome the fluctuation of customer demand, Without levelling, this fluctuation leads to underutilized capacities such as man and machine and specially the business and IT misalignment.

- Continuous improvement: follow the PDCA cycle, to create value, which must be a response to a need expressed by the costumer and must flow throughout the entire logistics chain of the company, in this case, the use of lean tools can help IT professionals better understand the value of their information system, with key performance indicators to define the value by referring to the users point of view and not the computer designer.

Identifying the value sought by the customer (easy navigation, confidentiality of data, good interface....) is usually a difficult task to do, as an example when the customer wants to make purchases online, the purchase process is much simpler when the site just asks the address and the payment method of the client, while other site asks to create an account and to activate this account by clicking on a link sent by email before being able to order. This complexity makes the customer no longer recommend on the same site. Lean can help to identify the value of customer by using the grid of analysis of the value elaborated by Daniel Jones and James Womack which define three types of activities in the value stream:

- Value-Added: Those activities that unambiguously create value.
- Type One Muda: Activities that create no value but seem to be unavoidable with current technologies or production assets.
- Type Two Muda: Activities that create no value and are immediately avoidable [7].

To sum up, information system using lean tools recommend that:

- Reports and other outputs from the system should only be produced if they add value and if they are useful to the decision makers and that they should only be sent to those who need them;
- Information should be processed quickly so that users do not have to wait for it;
- Continuous improvement: the providers and users of information should meet regularly to review the usefulness of existing information and identify improvements;
- Information systems should be flexible enough to meet special ad hoc needs or changing needs of managers over time. An information system that can only produce a standard set of reports is not lean. A system that allows managers to create their own customized reports from databases is more likely to be Lean.

5.3   Challenges of the Application of Lean in IT

Lean IT Still Has Challenges for Value-Stream Visualization, Unlike lean production, which is the base of Lean IT principles, it relies on digital and intangible rather than physical and tangible streams of value, these flows can be difficult to quantify and visualize, another challenge that companies may face when they use lean in their information system is related to cultural change, computer scientists always tend to work with their technical knowledge and do not care to integrate the relationship with the profession as a major axis of development of skills.

The implementation of Lean thinking can be broken by several elements such as human conflicts within the teams, the resistance to change, the principle "think product, not project" which is very disturbing for a team of developers who have not been trained and accustomed to this. The customer who has become more and more demanding and who does not know what he wants. As result, the program becomes more complex and contains several iterations; therefore the code will be insufficiently mastered.

This presents a challenge in terms of determining the lean practices that are applicable to IT support service environment and in devising ways to integrate these lean practices with process improvement framework like CMMI.

6   Conclusion

This paper discusses the development of a new approach that was initially applied to the industrial world to support the improvement and productivity of the information system. It is argued that there are many tools and methods of IT governance to improve some aspects of IS, but none of these methods allow the company to measure the robustness of its information system or to know and eliminate the non-value added activities. Nobody can deny the great revolution achieved through the application of lean in the industrial sector. The basic proposition of this article is that this tool could also be applied to the service functions and in particular to the IS where information management can be considered the product to be improved in order to meet the needs of the end user.

But, Lean IT Still Has Challenges for its application. These problems will be the subject of a future paper, we will address the obstacles that business can encounter by introducing lean into its information systems, through a case study and we will define how to move from the methods used in the industrial world to those used in the information systems to successfully implement the lean IT management tools.

Acknowledgement. At the end of this Paper, we express our gratitude and our sincere thanks to all those who contributed directly or indirectly to the realization of this article in particular M. Cherti and M. Bahaj who assisted us during the various stages of our work. Their guidance and valuable advice and availability have allowed us to master and overcome the challenges of this communication.

References

1. Elidrissi, D., Elidrissi, A.: Contribution des systèmes d'information à la performance des organisations: le cas des banques. la Revue des Sciences de Gestion **241**(1), 55–61 (2010)
2. Laudon, K.C., Laudon, J.P.: Management Information System, Managing the Digital Firm, 12th edn, pp. 41–50. Prentice Hall, Upper Saddle River (2012)
3. Emiliani, M.L.: Origins of lean management in America: the role of Connecticut businesses. J. Manag. Hist. **12**(2), 167–184 (2006)
4. Grzelczak, A., Werner-Lewandowska, K.: Importance of lean management in a contemporary enterprise – research results. Res. Logist. Prod. **6**(3), 195–206 (2016)
5. Schwalbe, K.: Information Technology Project Management, 8th edn. Cengage Learning, Boston (2015)
6. Laudon, K.C., Laudon, J.P.: Management Information Systems: Managing the Digital Firm, 15th edn. Pearson, London (2018)
7. Womack, J.P., Jones, D.T., Roos, D.: The Machine That Changed the World. Free Press, New York (1990). SKU: 9794
8. Khmelevsky, Y., Li, X., Madnick, S.: Software development using agile and scrum in distributed teams. In: 2017 Annual IEEE International Systems Conference (SysCon) (2017)
9. Kheirandish, M.R.F., Khodashenas, H., Farkhondeh, K., Ebrahimi, F., Besharatifard, A.: An analysis on the evolution of Management Information Systems (MIS) and their new approaches. Interdisc. J. Contemp. Res. Bus. **4**(12), 491–495 (2013)
10. Betz, C.T.: Architecture and Patterns for IT Service Management, Resource Planning, and Governance: Making Shoes for the Cobbler's Children, pp. 4–5. Morgan Kaufmann Publishers, Elsevier, San Francisco (2007)
11. Ignace, M.-P., Ignace, C., Médina, R., Contal, A.: la practique du Lean management dans l'IT. Pearson, London (2012)
12. Brand, K.: IT governance based on COBIT 4.1-A management Guide. Van Haren Publishing, Zaltbommel (2008)

Intrusion Detection Systems: To an Optimal Hybrid Intrusion Detection System

El Mostafa Rajaallah[1(✉)], Samir Achraf Chamkar[2],
and Soumiya Ain El Hayat[3]

[1] Laboratory: Security's Dynamics, Institute of Sport Sciences,
Hassan 1st University, Settat, Morocco
rajaallahelmostafa@gmail.com

[2] National School of Applied Sciences, Ibn Tofail University, Kenitra, Morocco
samirachrafchamkar2@gmail.com

[3] LITEN Laboratory, Faculty of Science and Technology, Hassan 1st University,
Settat, Morocco
Soumya.ainelhayat@gmail.com

Abstract. New information and communication technologies are increasingly dominating our daily lives to the point of becoming indispensable. The computerization of society is now a reality, no sector is aloof from this transformation. This human invention, which has made progress, certainly facilitated the lives of human beings, but at the same time generated new illegal behaviors: computer crimes. Cybercrime has a very broad meaning, characterized by dematerialization, anonymity, internationality, the technicality of matter and its rapid evolution. Cybercrime refers to the use of digital and electronic capabilities, or software to destroy, eliminate, and exploit public or private information systems, there are several definitions of cyber-security; this concept does not yet have a globally accepted definition. The various attempts at definition have shown how this notion is vast, complex and affects many fields; also this concept can be seen as set of computer processes aiming to protect data transiting over the Internet. The intrusion detection systems (IDSs) and intrusion prevention system (IPSs) play an essential role in detecting anomalies and attacks in the networks or system and to protect data transiting. In this paper, we survey IDS open sources whose objective is to propose a hybrid IDS.

Keywords: IDS · IPS · Cybercrime · Cyber-security · Hybrid-IDS

1 Introduction

To confront malicious actions against information systems, modern information systems include security engines that implement the adopted security policy. This is done by using technical tools of protection which include hardware software and cryptography protection. For the reason to make attacking the system more difficult, detect the occurrence of attacks, and take countermeasures to get rid of the consequences of an attack. Among these tools, Firewalls, Anti-virus, and Access control, etc. They implement the adopted security policy and grant the CAI (confidentiality, Availability,

© Springer Nature Switzerland AG 2019
F. Khoukhi et al. (Eds.): AIT2S 2018, LNNS 66, pp. 284–296, 2019.
https://doi.org/10.1007/978-3-030-11914-0_30

Integrity) of the system. With the help of these tools, each user has permissions that define if he can be permitted or prohibited, in order to access information stored in the computer, or to remote access via communication link on other computers. However, these tools of differentiation of local access cannot protect the system against malicious actions committed by an offender, which is permitted to access the resources. Such attacks are based on vulnerabilities discovered in software-hardware server and desktop station of IS. This is where IDS can take place to face both internal and external network attacks.

An Intrusion Detection System (IDS) can monitor operating system events, network traffic or file system status, to find evidence about the occurrence of attack and it reports the malicious actions to the security administrator. Therefore, IDS uses three detection methods, misuse detection that use signatures or patterns to identify attacks and anomaly-based detection, which are focused on the deviation of the normal behavior and raise an alarm in case of deviation. Hybrid detection combines both misuse detection and anomaly-based detection.

The remainder of this paper is organized as follows. In the first section, we present the important role of intrusion detection systems, then we make a taxonomy of IDSs and we explain how misuse detection and anomaly-based detection works by shedding the light on the advantages and the drawbacks of each method. In the second section, we survey (snort, Suricata, BRO) open source network IDSs to show their characteristics, and their ability to detect malicious actions. In the third section, we present the most used evasion attacks to evade misuse based detection systems such as snort IDS. In addition, we focus on shellcode mutation attack that changes the signature of the malicious code. In the last part, we make a comparison of PHAD, NETAD and ALAD anomaly based statistical algorithms, which monitor anomalous attributes in packets, and identify unusual actions on the computer network.

2   Related Works

Most open source intrusion detection systems such as Snort [8], Bro [3] and Suricata [1] use patterns or signature of attacks to protect the system, by comparing the incoming traffic with the available patterns of attacks included in a database, when a match occurs, IDS notify the security administrator about the intrusion or the suspicious behavior.

The process of detecting malicious actions masked by evasion techniques [6] is a difficult task for signature-based Intrusion Detection Systems. There are several techniques such as denial-of-service, Packet splitting, duplicate insertion, and shellcode mutation transforms the attacker's shellcode to escape signature detection.

Anomaly detection such as PHAD [4], NETAD [9] and ALAD [10] learn about the normal activities of the network by using a statistical model and raise an alarm in case of observing abnormal activity. Lower probability of an event leads to a higher anomaly score. PHAD, NETAD, and ALAD differ at the number of attributes and the attributes that they monitor. PHAD (Packet Header Anomaly Detection) has 33 attributes, corresponding to the Ethernet, IP, TCP, UDP and ICMP header fields. ALAD (Application Layered Anomaly Detection) models incoming server TCP requests,

NETAD model the first 48 attributes, consisting of the first 48 bytes of the packet starting with the IP header.

Aydin et al. [7] proposed a hybrid intrusion detection system which combines snort as misuse detection based and PHAD and NETAD as anomaly-based detection. They test their system on IDEVAL dataset, snort detected 27 attacks out of 201 and when they added PHAD and NETAD, the number of detected attacks increased up to 146 of 201 attacks.

3   Intrusion Detection System

Intrusion Detection System or IDS are network monitoring systems that have become almost indispensable due to the incessant increase in number and danger of network attacks in recent years.

3.1   Intrusion Detection System Taxonomy

The following figure shows Taxonomy of intrusion detection systems (Fig. 1).

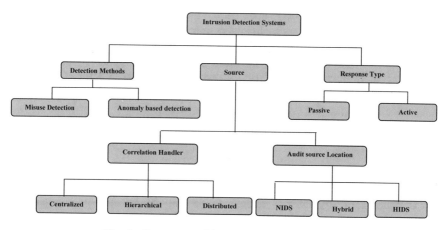

Fig. 1.   Taxonomy of intrusion detection systems

IDSs classified depending on the monitoring layer, Detection Layer, Alert Processing Layer, Reaction Layer. For monitoring, detecting, analyzing, and responding to unauthorized activities. Monitoring layer.

3.2   Data Collection

IDS can be used individually where each IDS works in each node independently or in combination with other IDSs. Collaborative IDSs consists of multiple IDSs over a large network where each one communicates with others. Each IDS has two main functional components: detection element and correlation handler. There are three structures of communication.

3.2.1   Central Structure

Each IDS acts as a detection element where it produces alerts locally, the generated alerts will be sent to a central server that plays the role of a correlation handler to analyze them. Through a centralized management control, an accurate detection decision can be made based on all the available alerts information.

3.2.2   Hierarchical Structure

The whole system is divided into several small groups based on similar features and similar software platforms. The IDPSs in the lowest level work as detection elements. on the other hand, the IDPSs in the higher level is furnished with both a detection element and a correlation handler, and correlate alerts from both their own level and lower level. The correlated alerts are then passed to a higher level for further analysis.

3.2.3   Fully Distributed Structure

There is no centralized coordinator to process the information; it compromises fully autonomous systems with distributed management control. All participant IDPSs have their own two main function components communicating with each other.

3.3   Audit Source Location

The nature of IDS depends also on the source of information used in the process of intrusion detection.

3.3.1   Network Intrusion Detection Systems

The NIDS are placed at strategic points within the network to monitor traffic. They perform the analysis traffic on the entire subnet and match the traffic passing through it to the library of known attacks. Once an attack is identified, or abnormal behavior is sensed, an alert is generated and sent to the security administrator.

3.3.2   Host Intrusion Detection Systems

The HIDS run on independent hosts or devices to monitor the state of a computer system. HIDS are able to check file accesses, changes to file permissions and which program accesses what resources. If the system files were modified or deleted, an alert is sent to the administrator.

3.3.3   Hybrid Intrusion Detection Systems

The Hybrid intrusion detection systems Combine information from a number of sensors. Often, both Host and Network-based intrusion detection systems in a central analyzer that is better able to identify and respond to intrusion activity.

3.4   Time of Detection

3.4.1   Real-Time Detection

IDS runs in the background, checking the files and network traffic depending on the type of IDS (HIDS or NIDS), comparing it to known viruses, worms, and other types of malware.

3.4.2   Non Real-Time Detection

It can fill the network inherent security gaps associated with vulnerability, to various types of attacks (especially DOS) that are not detectable by a common approach of audit trail analysis, and it has less resource consumption.

3.5   Reaction Layer

3.5.1   Passive System

The IDS notifies the security administrator by generating alerts and make a detailed report about the experienced malicious action. Furthermore, it generates a backup file in case of losing data.

3.5.2   Active System

The IPS (Intrusion Prevention System) can take countermeasures against experiencing an attack by terminating the malicious process and the session of the user, and by blocking the IP source of the attacker machine and disconnect it from the network.

3.6   Detection Methods

This is the most important part of this taxonomy; we have two main detection methods. First, misuse based detection rely on specific patterns to identify attacks. In contrast, anomaly detection refers to the use of a baseline of normal usage patterns, and whatever deviates from this gets flagged as possible intrusions.

3.6.1   Misuse Detection

Misuse or signature-based detectors analyze system activities, looking for events that match a predefined pattern of events describing a known attack these events are called signatures [2]. The system relies on a database that contains the signatures of all known attacks, and each time the system must interact with the database to make a comparison of the stored signatures with the incoming packets, and/or command sequences to assure that the traffic is clean at least from the attacks stored in the database. This technique represents knowledge about the bad or unacceptable behavior and seeks to detect it directly.

Advantages of Misuse Detection

- Misuse detection is very efficient in detecting attack without signaling too many false alarms. (The patterns of attacks must be well defined by a security administrator).
- Misuse IDS can help users and administrators to know and monitor their systems. Even if they are not security experts because each signature is related to an attack.

Then, any match of a signature or pattern with incoming packets gives the administrator the exact information about the attack so that he can deal with it quickly.

- The Misuse detection can be deployed without knowing the network, and it doesn't take time to learn about users and behaviors and data. The rules are defined from the beginning and updated each time they are available the Misuse detection can be deployed without knowing the network, and it doesn't take time to learn about users and behaviors and data. The rules are defined from the beginning and updated each time they are available.

Disadvantages of Misuse Detection

- The most negative point of this method of detection is that it can only detect known attacks, which the patterns or signatures are defined in the database. For this reason, the system needs to be updated each time with the newly discovered signatures.
- The misuse detection can Miss Zero-day attacks which is undiscovered software hole that could be exploited to affect hardware, application, data. In simple word, it refers to be undiscovered vulnerabilities that are unknown to the user and has no patch or fix yet.
- When a well-known attack is changed slightly and a variant of attack is obtained, the IDS will be unable to detect this new variant of the same attack.

3.6.2   Anomaly-Based Detection

An anomaly-based detection system designed establishes a baseline of normal usage patterns, and flags anything else as possible intrusion [5]. Anomaly detection generates an alarm in case of new events not observed during training that occurs with a frequency greater or less than two standard deviations from the statistical norm.

Advantages of Anomaly Detection

- Anomaly-based detection can detect new attack which the signature is not defined in the database and even when information about the attack does not exist.
- IDS adapt it to be more accurate by the time.
- No need to update the signature database to detect new threats.
- Very little maintenance once the system is installed it continues to learn about network activity and continues to build its profiles.
- Disadvantages of anomaly-based detection:
- Anomaly-based IDS generally flags many false alarms, because deviating from normal behaviour does not always mean that an attack is occurring.
- The anomaly-based approach requires a large set of training data that consist of system event logs in order to construct normal behaviour profile.
- Require time to be ready to be used with the real data to protect the system.
- Can be used by a malicious person in the training period to make the malicious actions normal for the IDS, so that it can evade detection.
- The high percent of false alarms that are typically generated in anomaly detection systems make it very difficult to associate specific alarms with the events that

triggered them. It requires security administrator skills to protect the system and to make a separation between the real attacks and the false alarms. It is too difficult to discover the boundaries between abnormal and normal behaviour.

4 Comparison of Open Source IDSs

Open source IDSs is a good option for companies which do not have as much money as the larger companies and governmental organizations have. When choosing the open source intrusion detection system, the security or network administrator should have prior knowledge about them.

It is very important to have previous knowledge about IDSs to make the choice. In our case, we make a comparison of Snort, SURICATA and Bro open source IDS and we make a comparison of them.

Snort and Bro IDS existed since 1998, while Suricata first in July 2010.

The choices you make as a leader will affect the quality of the network security. We made a comparison of Snort, Suricata, and Bro, depending on many features, such as CPU and memory consumption, rules, documentation, and installation, etc. (Tables 1, 2 and 3).

Table 1. Characteristics of BRO.

	Bro
Multi CPU	Bro provided a "worker"-based architecture to utilize multiple processors
IPS function	Bro provides only limited intrusion prevention functionality; Its scripts can execute arbitrary programs, which are used operationally to terminate misbehaving TCP connections
Support to high speed networks	high

Table 2. Characteristics of SURICATA.

	Suricata
Multi CPU	Suricata could be considered as an extension of Snort for large networks. Using multiple CPU
IPS function	Intrusion Prevention with NFQUEUE
Support to high speed networks	high

Table 3. Characteristics of Snort.

	Snort
Multi CPU	Snort is highly efficient in the scenario of moderate traffic with a single core processor
IPS function	Snort Inline support new rule keywords to allow traffic matching a rule to be dropped, thus turning Snort into an Intrusion Prevention System (IPS)
Support to high speed networks	medium

5   Evasion Techniques

The process of detecting attacks disguised by evasion techniques is a challenge for signature-based Intrusion Detection Systems and Intrusion Prevention Systems, and there are many evasion techniques can be used to evade recent signature detection systems, Such as Packet splitting, Denial-of-service, Payload, and shellcode mutation [6].

In this part, we focus on the effectiveness of shellcode mutation using some Kali Linux framework against Snort signature-based.

5.1   Shellcode Mutation

5.1.1   Shellcode

Shellcode is basically a list of crafted commands that can be executed once the code is injected into a running application. It's a series of instructions used as a payload when exploiting vulnerability. A command shell will be provided after the set of instructions have been performed by the target machine (Fig. 2).

Under normal conditions, the program follows the instructions made by the creators (programmers). A Hacker could make a program follow his orders and ignore the command creator, allowing the attacker to execute any code he wants.

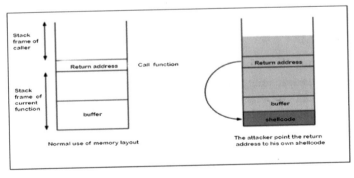

Fig. 2. Overwrite return address to points to the shellcode.

5.1.2 Polymorphic Shellcode

IDS checks incoming and outgoing packets. When a packet contains patterns considered as malicious, or the signature of the shellcode already exist in IDS database and reused. Then, the IDS will easily detect and prevent the attack. This is where Polymorphic shellcode takes place to get away from the detection by AV, and Intrusion Detection System by changing it into many kinds without altering the main function and the end result.

Since there are too many ways to mutate a shellcode, detecting polymorphic codes on an IDS is particularly tricky. IDS may need to decrypt the encrypted code to restore the original signature or even emulate the code execution.

To find malicious behavior. The tasks of restoring the shellcode semantics online are, therefore, computationally expensive and burden the load of an IPS.

5.2 Evasion Tools

There are many well-known free tools that can be used to evade Intrusion detection systems and Antivirus systems, by exploiting TCP/IP protocols, transform URI requests, transform the shellcode and provide many other functionalities, among these tools.

5.2.1 Metasploit [11]

This is an open source project which provides resources for discovering vulnerabilities and security issues. It allows a network administrator to test the security level of his own network, and identify the security risks. Furthermore, it can be used to develop exploits codes against a remote target host, build malware signatures and execute penetration testing. Moreover, it provides a number of polymorphic shellcode encoders that could be used, to change malware signature to bypass Antivirus and IDS detection, and can establish a connection with the target host in case of completion of attack.

5.2.2 Msfvenom

Msfvenom is an open source penetration testing tool which combines two other tools. Msfpayload that allow you to create a payload depending on the target system and Msfencode allow you to evade detection using deferent encoders, putting both of these tools into a single Framework instance as of June 8th, 2015.

5.2.3 Veil-evasion

This tool affords the possibility to create and transform the malicious scripts and hides it within another file, converts them to exe, and encrypt them. Moreover, Veil-evasion affords multiple programming languages, different delivery options, obfuscation functions, and Metasploit payload integration.

6 Anomaly-Based Detection Using PHAD, NETAD and ALAD Statistical Algorithms

Most network anomaly detection systems are designed to make a difference between authorized and unauthorized users by checking the network traffic and compare it with the habitual traffic.

Using the difference between hostile and benign traffic in ways that anomaly traffic can be distinguished without knowing the nature of the attack, NETAD also classifies novel events as anomalous if they are sufficiently rare and have not occurred recently.

6.1 PHAD (Packet Header Anomaly Detection)

PHAD models 33 attributes which correspond to packet header fields with 1 to 4 bytes. The Fields smaller than one byte are combined into one byte. Fields larger than 4 bytes (such as 6 byte Ethernet addresses) are split. The attributes are as follows:

- Ethernet header: packet size, source address, destination address (high and low 3 bytes), and protocol.
- IP header: header length, TOS, packet size, IP fragment ID, IP flags and pointer (as a 2 byte attribute), TTL, protocol, checksum (computed), and source and destination addresses.
- TCP header: source and destination ports, sequence and acknowledgment numbers, header, length, flags, window size, checksum (computed), urgent pointer, and options (4 bytes if present).
- UDP header: source and destination ports, checksum (computed), and length.
- ICMP header: type, code, and checksum (computed).

6.2 NETAD (Network Traffic Anomaly Detection)

NETAD models 48 attributes that contain the first 48 bytes of each packet starting with the IP header. For a TCP packet, NETAD will deal with the 20 bytes of the IP header, and other 20 bytes of TCP header with 8 bytes of the application payload.

NETAD uses two steps in the detection process starting with:

Traffic filtering stage : The most attack used against a target server or operating system can be detected by checking only the first few packets of incoming service requests.

The modeling stage of NETAD models nine types of packets, for a total of - 9 x 48 = 432 rules. The nine models represent commonly used (and commonly exploited) protocols. The rules were selected because they give good results experimentally.

6.3 ALAD (Application Layered Anomaly Detection)

ALAD anomaly detection a score to an incoming server TCP connection, and it is configured to know the range of IP addresses. It is supposed to protect, and it distinguishes server ports (0-1023) from client ports (1024-65535). This model considers that most attacks are initiated by the attacker (rather than by waiting for a victim), and are, therefore, against servers rather than clients. This model is based on the combination of five attributes that give the best performance that gives a high detection rate at a fixed false alarm rate.

7   Comparison of PHAD, NETAD, and ALAD

We will obtain the results mentioned in the following tables, by using the 1999 DARPA (Defense Advanced Research Project Agency) Intrusion Detection Evaluation: an off-line evaluation and a real-time evaluation [11, 12] (Tables 4, 5 and 6).

Table 4. Performances of PHAD anomaly-based statistical algorithm.

	PHAD
Implementation	Easy to implement (400 lines of C++ code)
Traffic filtering	- No - Examine all traffic
Number of detected attacks in 100 false alarm out of 201 attacks	27
Total number of detected attacks out of 201	54

Table 5. Performances of NETAD anomaly-based statistical detection.

	NETAD
Implementation	~ 300 line C++ program The filtering program is another ~ 250 lines of C++
Traffic filtering	- Yes - Only start of incoming server request are examined
Number of models	models 9 subsets of the filtered traffic corresponding to 9 common packet types
Number of detected attacks in 100 false alarm out of 201 attacks	132
Total number of detected attacks out of 201	152

Table 6. Performances of ALAD anomaly-based statistical algorithm.

	ALAD
Implementation	A 400 line C++ program to reassemble TCP packets into streams, and a 90 line Perl script to analyze them
Traffic filtering	- Yes - Incoming server TCP connection
Number of detected attacks in 100 false alarm out of 201 attacks.	45
Total number of detected attacks out of 201	60

8 Conclusion

Depending on the results of evasion techniques and the comparison of open source IDS, signature-based systems such as Snort are really powerful to detect known attacks, but not sufficient to stop targeted attacks. Especially, if you only use the default Snort rules, since the attacker can test their attack in advance to avoid detection.

The use of anomaly-based detection (PHAD, NETAD and ALAD) can detect unusual events in a vast amount of data, and brings them to the attention of a network security expert for analysis.

We will be able to increase the number of detected attacks, by combining misuse detection (detect known attacks) and anomaly detection (detect zero-day attacks) in a hybrid model.

References

1. Suricata: Suricata Open-Source IDS/ IPS/ NSM engine (2017). https://suricata-ids.org
2. Boiko, A., Shendryk, V.: System Integration and Security of Information Systems. Procedia Comput. Sci. **104**, 35–42 (2017)
3. Sommer, R.: The Bro Network Security Monitor. Bro.Org. 1 (2016)
4. Garg, A., Maheshwari, P.: PHAD: packet header anomaly detection. In: Proceedings of the 10th International Conference on Intelligent Systems and Control, ISCO 2016 (2016)
5. Nadiammai, G.V., Hemalatha, M.: Effective approach toward Intrusion Detection System using data mining techniques. Egypt. Inform. J. **15**(1), 37–50 (2014)
6. Cheng, T.H., Lin, Y.D., Lai, Y.C., Lin, P.C.: Evasion techniques: Sneaking through your intrusion detection/prevention systems. IEEE Commun. Surv. Tutorials **14**(4), 1011–1020 (2012)
7. Aydin, M.A., Zaim, A.H., Ceylan, K.G.: A hybrid intrusion detection system design for computer network security. Comput. Electr. Eng. **35**(3), 517–526 (2009)
8. Rehman, R.U.: Introduction to Intrusion Detection and Snort (2003)
9. Mahoney, M.V.: Network traffic anomaly detection based on packet bytes. In: Proceedings of the 2003 ACM Symposium Application Computing - SAC 2003, p. 346 (2003)

10. Mahoney, M.V.: a machine learning approach to detecting attacks by identifying anomalies in network traffic. Computer (Long. Beach. Calif.) (2003)
11. Lippmann, R., Haines, J.W., Fried, D.J., Korba, J., Das, K.: The 1999 DARPA off-line intrusion detection evaluation. Comput. Netw. **34**(4), 579–595 (2000)
12. Mahoney, M.: Source code for PHAD, ALAD, LERAD, NETAD, SAD, EVAL3, EVAL4, EVAL and AFIL.PL. http://cs.fit.edu/~mmahoney/dist/

Author Index

© Springer Nature Switzerland AG 2019
F. Khoukhi et al. (Eds.): AIT2S 2018, LNNS 66, pp. 297–298, 2019.
https://doi.org/10.1007/978-3-030-11914-0

Printed in the United States
By Bookmasters